Simulation Theory

A Reference Book for Simulation Theory

(A Psychological and Philosophical Consideration)

Maria Baker

Published By **Simon Dough**

Maria Baker

*Simulation Theory: A Reference Book for
Simulation Theory (A Psychological and
Philosophical Consideration)*

ISBN   978-1-77485-717-5

No part of this guidebook shall be reproduced in any form without permission in writing from the publisher except in the case of brief quotations embodied in critical articles or reviews.

Legal & Disclaimer

TABLE OF CONTENTS

Chapter 1: Bostrom's Theory

Do you live in a computer-generated simulation? Based on Nick Bostrom, a Swedish-born professor at the University of Oxford in England--a winner of the Professorial Distinction award having a degree in the field of theoretical Physics, computational neuroscience logic, logic, as well as artificial intelligence -- you and everyone else on Earth most likely are. However, like Neo in the film, The Matrix, you just don't realize that.

When Bostrom published his 2003 paper which suggested there's a good possibility that our world is a computer-generated simulation, Bostrom caused shock waves across the world of academia. The concept is now a commonplace in popular culture. Some people believe it's possible that the world we live in are the result of a computer-generated simulation developed by a sophisticated civilization. As of 2016, as an instance an article in the New Yorker article reported that the majority of people living within Silicon Valley had become obsessed with the idea and two billionaires who were not named have hired teams of scientists to get out of the technology of simulation. Business mega-millionaire Elon Musk has come out and admitted his own fascination with the idea, as well as his belief that the possibility of living in the real world, i.e., one which is not simulated is one in billion. A lot of prominent scientists, prominent people and philosophers are now believed to are taking seriously the idea that humans, as well as our world, are created by the computer that is a gigantic one because they believe it's only an issue of time before we can have the technology to create a digitally created real world. Over the last fifty years or so humans have constructed ever more realistic virtual worlds. Simulation theory advocates say that currently nothing stands between these simulation realities advancing in infinitely. Many say they see that in the future, these virtual worlds will be real enough that, even if we do not have direct access to of the virtual world, it will be difficult to distinguish the different between a virtual environment and reality. This begs the concern, "How do we know that we're not living in a virtual world?" Could this have already occurred? Perhaps we are living the realm of what Bostrom refers to as "an ancestral simulation," created by an advanced civilization that was created to study its own history? Bostrom suggests that advanced civilizations have numerous reasons for running simulations, and they could generate massive amounts of them. This would logically be logical that the vast majority of conscious beings that could exist in the universe are in or reside in simulations.

Bostrom's argument is valid I believe that, if it weren't due to a flawed base the thesis could be valid. In fact, it appears like in a few ways, it's near to what I believe to be is the truth about reality in the physical realm and that's the main reason that many smart and thoughtful people find it compelling. It is evident that a lot of the information Bostrom states is true for their own. But, as I'll discuss as you read Bostrom's theory is far from the mark.

If you've gone through the two books of the Life After Death series and you have considered the evidence offered in them thoughtfully and with great care You'll likely like me and reach the similar conclusion. This volume I'll provide information that is not in those books. I will try to avoid referring to evidence from the first two briefs in order to keep readers from becoming bored by these volumes.

Before I get started I'll an overview of the ideas Bostrom has offered. Bostrom lays out three arguments and argues that each of them has to be valid.

Proposition One states the idea that every technological society will die or die before attaining technological maturity. thus, we are not living in a virtual universe since, as a consequence that they don't exist.

* Proposition Two is that civilizations are able to attain technological maturity, but for reasons unknown, they do not develop simulations that are similar to our own world.

The third proposition Three is that if do not accept the first two propositions and accept his third and last claim that we are definitely living in a virtual world.

Let's look closely at Bostrom's proposals.

Proposition One

The primary argument is that civilizations cannot attain technological maturity, consequently, they don't get the chance to develop what Bostrom refers to as "ancestor simulations." It seems to be logical that if other civilizations like ours existed and exist, at the very least, some of them would eventually self-destruct. The tale of Atlantis is a good example. In our own situation we've developed nuclear weapons that give us the ability to destroy civilization. It's plausible that this could occur. Because of their theological views and their hatred of Jews and Jews, the Iranian mullahs appear to be intent on bringing about Armageddon. Let's hope that they don't ever become the next target for the bomb.

The logic seems to suggest that, the existence of civilizations is fairly widespread across the universe and are not uncommon, it is likely that each one of them would eventually self-destruct. We are part of a galaxy consisting of hundreds of billions galaxies, and they contain millions of systems of stars each of which could be the site of a growing technological civilization. What is stopping any of them from progressing to the technological maturity needed to build an authentic simulation?

I believe that the recent flurry of UFO sightings are a major argument against this theory However, certain scientists believe that there is there is no intelligent life on the universe beyond our personal. The argument of this viewpoint stems from what is known as the Fermi Paradox, which is that the universe around us is strangely silent. Therefore it is likely to be empty of life that is not the ones on Earth. However, according to many scientists that this universe is inhabited for the longest time starting eons prior to when our solar system was created. They claim that the universe has been going on since 13.8 billion years, and that Earth is just 4.6 billion years old. If this is true, it would appear that enough time has passed for advanced civilizations to emerge other than our own. Yet, we have no evidence of other civilizations to our current.

What is the reason for that? Certain scientists believe that the evolutionary steps between inert matter and living things are rare that they hinder existence from ever developing or even living things that develop into a state of consciousness.

Think about this: In 1953 James Watson [born 1928] and Francis Crick [1916-2004] discovered the double helix, also known as the twisting-ladder structure of deoxyribonucleic acids (DNA) which is essential for the existence of life. It was a turning point in the development of science and led to the development of modern molecular biology that is focused on knowing how genes regulate the chemical reactions within cells.

A few years later, in the year 1957 Francis Crick realized that the chemical subunits that line the inside of the double helix function similar to alphabetic characters found in the written language or digital characters like the zeros and ones found in computer code. He also realized that they were the ones controlling the production of proteins as well as proteins that cells require to survive. Also, digital information guides the creation of essential components in living cells. To explain the basis of life, it is necessary to describe how this complex processing process came into existence.

A lot of scientists, including Nick Bostrom, continue to remain steadfast to the basic idea that is the basis of Scientific

Materialism, which is that only physical matter, i.e., "matter," exists. Also, if it can't be observed under microscopes, it can't be. If this is the case the concept of consciousness and intelligence could be created before evolution created the brain. A lot of scientists believe in this idea, even over the years of intensive effort , they've not been able to pinpoint what the brain i.e. an inert substance, can achieve this.

This has led to a conundrum that an academic at NYU famous Australian psychologist and philosopher David John Chalmers, has called "The hard Problem." The problem was discussed in the 2014. Ted Talk, Chalmers proposed the theory of panpsychism that suggests consciousness could be a key element of the universe. It could be a fundamental building block like time, space and mass. A subatomic particle could possess a small amount of consciousness. He further stated that the degree of consciousness the individual is dependent on the volume of information being processed. That's why the human being has an elevated level of consciousness than dogs, an animal or even the steel piece. Chalmers's theory confirms Bostrom's hypothesis the sense that a computer that processes huge quantities of data would be aware. Furthermore, Bostrom maintains that a simulation's software can create consciousness for those who live within it. This is an important point to consider as we progress through this discussion.

The bottom line is: If Scientific Materialist theory is correct and even with the inclusion of panpsychism DNA would have to have been discovered through accident. If it was it could have happened even in the face of extremely long odds due to the fact that DNA is compared to an algorithm for computers that is superior to anything human beings have created.

How complex and advanced is it? In a report published on the web site for BBC Science Focus Magazine, the UK's most renowned science and technology magazine: "The DNA in your cells is packed into 46 chromosomes inside the nucleus. Additionally, since it is naturally helical DNA is super-coiled by enzymes, which means it takes less space. In the event that you extended the DNA of one cell to the maximum it would measure about two meters in length, and all the DNA contained in your cells roughly double the size of Solar System."

What's that amazing? Consider the vast quantity of information that is contained within DNA. Every cell of your body has six and one-half meters of information that is so small that you need an microscope to look it up!

Mathematics experts who calculated the probability of DNA occurring by chance claim that the odds are one in infinite. However in a universe that is literally filled with billions of solar systems, Bostrom believes that it's quite possible that life may have developed on other planets in addition Earth.

If life only occurred only once during 14.8 billion years because of the extremely long odds it is possible to argue that the sheer complexity of DNA suggests that we are living in a virtual world. It is highly probable that some form of intelligence was behind the origin of life in the way we see it. If it is believed, according to what Scientific Materials believe, that creator isn't God or any other intelligent source that is not material is the case, it could represent an intelligent civilization which was created accidentally somewhere else and we live in a model that the civilization built. It is possible that we live in a model created by our descendents, who have advanced in technology and running an ancestral simulation. Or we could live in a computer-generated simulation that Artificial Intelligence (AI) computers that run wild that our ancestors created, like the The Matrix films.

Proposition Two

The other argument is that civilizations may achieve technological maturity but despite this they do not create simulations that are comparable to ours. Simulations can serve as sophisticated laboratories that are able to answer questions that are unimaginable to minds at the moment of technological development. In many ways, simulations could be the most effective test of theories since the designers could have complete control over every aspect of the environment. With the immense potential of these simulations, what possible motives could stop our descendants from developing simulations? It is possible that the simulations which accurately like our own world aren't possible. In light of the increasing realisticity of our simulation environments it seems unlikely. So , why should we be concerned?

There are at most two possible reasons for why authentic simulations of ancestors may not be possible. The first reason is that attempting to simulate an environment like ours is computationally impossible since modeling every single of detail, from subatomic particles to massive galaxies would be cost-intensive that it wouldn't be feasible even in the most technologically advanced society. Michio Kaku, a prominent physicist, rejects the idea of simulation because of this. He believes that the only information processor that can stand capable of modeling a universe would be the Universe itself. This could or might not be the case however, it is not taking into account the fact that a convincing simulation does not require rendering each and every detail that exists in our universe. distant galaxies in view, as an instance can be reduced to a small amount of data, similar to in video games, which render the objects and scenes only when the avatar interacts with them. In this manner it is possible to drastically reduce computational costs by focusing on the parts in the model that have directly relevant at any point in. Additionally, memories can be incorporated into the people living within it, to give an illusion of continuity. So should the architects wish to create a world which is identical to the world we live in it would be necessary modify the experiences of the people who live there.

This is the second reason that the simulations aren't real. As in The Matrix films, the people living in the Matrix remain linked to physical bodies that remain in the real world. But, as we've already mentioned, Bostrom postulates that conscious people are a part of this simulation. There are many reasons to believe that replicating the mind of a person digitally isn't possible. We'll revisit this idea in the future, but if we continue with Bostrom's belief that the simulations that support consciousness exist, then we should think about the reasons why a modern civilization could choose to not create these simulations. The main reason for their denial could be based on ethical issues related to the treatment of conscious beings.

The world is not easy for all people in the world you and I live in the present. Certain people suffer immense pain and suffer. Do highly intelligent beings of an advanced civilization be willing to inflict the suffering of conscious, yet less-developed beings? Would they permit them to endure pain and then disappear, with no any benefit or compensation in any way to the person who suffered? They could. For instance, up until they were declared endangered during 2015 we sprayed our closest relatives within the realm of animals, the chimps, with diseases to examine the effects of drugs and medical methods. We do the same with other living beings, such as baboons, cats ferrets, dogs Hamsters, guinea pigs horses, llamas and mice. We argue that the pain we cause is justified since what we have learned is beneficial for us. On the basis of similar reasoning humans slaughter cattle, pigs, sheep and other animals to consume them. However, maybe at the point when sophisticated simulations are feasible, we will have to established laws that will prevent the kind of thing that happens.

If, on the other hand, there are numerous advanced civilizations in the universe according to Bostrom suggests, would each one of them have these rules? Most likely not. It will only take an advanced society to become engaged in the business of developing simulations to have an overwhelming number in existence at any time.

Proposition Three

Bostrom's third assertion is that we live in an imaginary world. He seems to believe that at least each of the three assertions is true , and chances are approximately equally for every. Elon Musk believes the chance that we're not in a simulation is just one in billion, so it is his duty to prove that we live in a simulation.

I think that, in a sense, it's true that we're living in a simulation however, it's not the kind of simulation Bostrom and Musk are thinking of. It is at least one thing that the simulation is that it's very different.

Chapter 2: Simulation Theory

When I heard about Simulation Theory, I was fascinated and decided to study it to find out what it's all about.

Like the majority of videos on YouTube as well as articles on the internet the same thing happens every time. One thing will lead to another. When I was learning about Simulation Theory, something would occur that I was unsure of and I'd have to investigate that.

However, the same thing happened repeatedly until I completed enough watching videos, and had read sufficient to return to the beginning and study Simulation Theory again and this time, I understood it.

In order to get there, I spent hours watching video and read articles after article about a variety of topics like simulation theory Hologram theory, what is gravity and how it affects space and time, Black holes, Dark matter, Particle accelerators, Particle Physics The double slit test, String theory, Quantum theory, Quantum physics, Mathematics and infinity Quantum Entanglement as well as Quantum computing.

From a simplistic perspective I found it interesting and decided to write some of it down, and then added my thoughts towards the end to make it easier to attempt to comprehend it.

My search began by examining my understanding of the Simulation Theory, that we all reside in a computerized simulation that we created at some point in the future when technological advancements would lead to computers that run simulations so realistic that we won't be able to distinguish the real world from them and that in any simulation, thousands of other people might be operating a multitude of simulations, and so on and on. It would mean that there could be literally millions or trillions of simulators being run simultaneously. Each simulation would have approximately 8 billion brains, that is 8 billion x trillions equals many.

The idea is that with this numerous brains, for us to be not in a simulation , and to be living in the real world it is a chance of one in billion. So, it's more likely than not to exist in a computer-generated simulation but don't realize the reality.

There are a majority of people tried to disprove this idea, but to date, they haven't succeeded. One approach was to examine the way computers work today in order to build an accurate simulation. It's claimed that all the parameters of what we believe to be real are the same as the ones that could be required in a computer simulation for example, the boundaries of our galaxy being impossible to reach and our planet as the only life on the planet. But a computer capable of running an application like that holds so much information that it makes it extremely unlikely.

In 1720, the question was asked the moment we shut our eyes, does everything exist beyond our minds. That set the science world on the road to find out the components of everything composed of. It was discovered that everything is made of atoms so tiny that they aren't visible by the naked eye however they comprise the entirety of what we see around us.

Then, these atoms were broken up and within were smaller particles. These electrons, neutrons, and protons were, after using particle accelerators dismantled to discover that there were even smaller quarks in. Science took a giant leap forward that brought many more inquiries than solutions, and new theories that need to be verified or disproved with the help of maths and science.

Take a moment to listen Everything I'm discussing is tied together to highlight an argument.

Mathematicians who were trying to prove and disprove new theories also came up with some of their own predictions and discoveries. One of them was about Crystals, and Quasicrystals. A crystal is because of the way in which it is held together. crystals that are tightly packed have the mathematical symmetry of three dimensions. If you put a crystal to the light and take a look at its shadow, you will see the appearance of a quasicrystal that is two-dimensional.

A shadow created by a 3D crystal turns flat and it becomes the 2-dimensional representation of an object that is 3D.

It was then realized that it is possible to see something in multiple dimensions was taken a step further by calculating the change between 3D in 2D it was also calculated that an 3D object can be mathematically attainable with 4D as well as 5/6, and the list goes the list goes on. It was later discovered that the mathematics of an eight-dimensional crystal was the best match for all maths about this, and it was appropriately called E8 which is also known as E8 Lattice.

Okay, now I can continue to investigate something that is awe-inspiring and unproved, but confirmed by the double-slit experiment.
Light waves, radio waves, and all those in the pond move in an outward-expanding direction however particles i.e. light particles are all traveling in straight lines. Imagine a light then a couple of feet from it, a wall that has two slits in the middle, and a few feet further back, the other solid wall.

When the light is turned on, you should think that you would see bright two lines appearing on the wall behind, but what you see are a number of lines visible on the back wall (an interference pattern or wave pattern) and this is a proof that light particles are traveling in waveform.

It was believed that this could be incorrect and that particles might be bounced off of each another when they passed through the slits. So they put particles through each one at a time and, after they had made it through the slits they left with an elongated wave. They wanted to find out what slit particles of the wave were passing through, so they set the camera that was aimed at the rear of the two slits . Then they conducted the experiment again however, the results were different. The second time, on the wall behind, there were two distinct glowing lines.

This discovery has proven that light is a waves until it is observed and then transforms into particle form , which alters the result. It was a mystery to scientists as it appears that particles or light waves move in a manner that suggests they are aware that they're being watched. This is consistent with the Simulation Theory that there is nothing in reality other than the waveform, until you take a examine it, as the simulation of a computer game functions.

A few years ago, there was a suggestion that this experiment be repeated, but this time using a delayed observation, putting cameras in the position right before the particles or light waves reach the back wall after they've passed between the two slits before deciding in the final seconds whether or not to watch the results. The results were quite shocking.

Time is now being questioned along with space.

What transpired was that the light that passed across the two openings in waves, and then at their point of view, it changed into particles, but it transformed backwards in time, back until they had not passed through the openings in the first place and as if they were particles from the beginning in the first place. Thus, by studying them and observing them, they changed the course of events prior to taking the decision to look them up. This has been proved, but cannot be explained.

This isn't the only particle-based experiment which can be proved but isn't explained.

Another instance is the case when two particles are at a point of entanglement and then are separated. This is called quantum entanglement. Or like Albert Einstein put it: "spooky actions from the distance" When two particles get entangled, they are able to move them away from one another, but they behave as two particles that are moving at the same time.

This test has been conducted over a long distance, however it is claimed that there is no distance between them can separate them once they are connected. With no possibility of sending an email across a huge distance in a single instant, which effectively eliminates the need for time and is still unsolved.

Let's look at what we are aware of about time. It is known that time travel is feasible when moving forward, because time is a function of to speed, perception, and mass. The bigger something is, and the more quickly it moves it is, the slower the time goes by in comparison to the perception of other.

The most straightforward explanation I've come across is:

If you're in a big space with a stage and an audience, you place an individual on a stool on the stage, facing the audience. Directly in front of the person, you set a lamp on a pendulum that represents the time and indicate two different points.

When the pendulum swings towards the left mark point A, and then swings towards points B to the left. When the person sitting in the chair looks at its light source, the pendulum is travelling through their eyes with the velocity of light, so you can take note of the amount of time it takes the pendulum to move through point A until B, using the reference eye's point of view. it. Now, at the same time, using the same pendulum, you can calculate the duration for it to travel from A point B, this time, from the viewpoint of the people standing 30 feet away.

We know how fast light moves, and can determine that it takes longer to reach the people farther away, so if we measure the distance between places A & B we know that, despite exactly the same length, it was more time that the crowd to go the distance as it did for the person sitting in the seat which means that the person sitting in the seat is moving through time quicker that the crowd.

But, the amount is so tiny that it will never make any impact.

Recently, the American female spaceflight astronaut came home to earth after spending 64 days of space travel around the earth. It was discovered that she grew 2.5 seconds more slowly than the rest of us.

If you were to travel into outer space and travel around the world in the fastest way possible, and then stay for 20 years, at the time you return you'd be 20 years old, but everybody else would be more than you, so you would have traveled further into the future. As we move closer to our speed the speed of time decreases. Therefore, if we were able to travel at speeds of light the 20 year space time would be one million years here on earth.

The notion of time travel going backwards has ever been confirmed, although there are numerous theories about black holes, wormholes or also time crystals.

My opinion lies in the fact that time travel isn't only possible , but is occurring all over the world. I was convinced of this when I heard about a recent invention that made use of quantum computers.

An Quantum computer is in development in the present. Our current computers the chips that power them utilize bits i.e. the 0s and 1s. The word 0 means 1, and one with the meaning of yes. with electricity, you can make it possible to open and close the string of 1s and 0s as closing and opening gates in a way which can perform extremely fast calculations. There are however limits to the amount of calculations you can do and how much power is required to create these calculations, as well as the time needed to process large amounts of data.

One example is when you have 16 people in the room and you were asked to group them in as many combinations as you could, such as John = 1, Peter = 2, Helen = 3, Helen & Peter = 4, and then on. At the point you reach the end you will have 2.4 trillion possible combinations. When you add the 17th, 18th, 1000 individuals, there's a threshold that it's unintelligible due to the power and time it will require a computer the various combinations.

However, as we've discovered particles may have multiple value they are in the same area by deciding to look at them or not. They could be put into the form of a computer chip.

Thus, if an element has a value of two, and the one next to it has a value of two, and the particle next to it has two values and it goes on The number of paths or strings increases exponentially. Like placing the rice grain on a square of chessboard and then x2 on the next square, and x2 in the next, and then on.

When you are at the 64th square , the amount of rice grains is (I was required to confirm the number) 18 quintillion, 4 hundred and forty-six quadrillion seventy-four trillion seventy-three billion seven nine million five hundred and fifty-one thousand six hundred and fifteen grain of rice.

It is in the stage of development and is referred to as the quantum processor, which will use for quantum computation. This breakthrough could revolutionize computing technology to levels that are far beyond what we imagined possible.

So, my conclusion:

Simulation Theory and all that is associated with it has brought me to discover what waves do when observed, which eventually led to the development of quantum processing, which could open the way to possibilities for Simulation Theory being real through the advancement of technology.

In other words, we are currently studying something that isn't yet in existence and this is the reason we're unable to comprehend it. If we don't move forward in order to understand the future, then nothing could be accomplished and we'd always be stuck in the present.

It is my belief that time forms the foundation of all things and that time exists as we're definitely moving through it, however our understanding that it is a linear line with a beginning, middle and a climax is where my beliefs end.

I don't see it as a line , and more importantly, since it is everywhere. I believe we are engulfed in time, but we don't be able to comprehend it. It's not an object, but a method to track our journey through life, or time is itself a state of being. It is how our brains decide where and how big an object is in relation to the place and time we are as well as how we travel from one place to the next with respect to all the other objects surrounding us, which is in motion or not.

(If we were to stand still, would time cease to exist? The answer is no because , if we were to stand still all the rest of the world continues to move. If the entire world stopped moving, there was no reason to need time.).

If we didn't have our perception of time, there is nothing to see. Therefore, time is the source of our existence and consciousness creates time. That means tomorrow, today , and yesterday are happening at the same time however, our brains are able to determine when and where we are at the moment.

I hope you understood that it was difficult to convey.

When it comes to Simulation Theory is concerned, I'm unable to think of a reason to believe it isn't true, and when all questions are answered, it seems like it is more likely it is. Are I convinced? yes, but I'm still willing to believe it, but until I can find actual evidence that I am able to say that it's possible in a world that is full of possibilities.

The main questions I'm faced with is that it's apparent to me, but until now I'm unable to find what the solution is.

"What occurs when you mix the Double Slit Experiment along with quantum Entanglement?"

If two particles interact like one in the distance, and you can observe one at this point to change shape of a wave into a particle What happens to the second entangled particle on the other side? If it changes, could this not be recorded and reversed creating a two-way communication across space and time?

OR

Since it's been demonstrated that particles reverse their direction with time, is it not logical to bounce a beam of particles off satellites and cover the globe many million times and then look at it in an Morse like pattern.

Do you know the numbers for tomorrow's lotto? You could you can save some people's lives by stopping criminals prior to them committing the crime.

I've recently had a discussion about what ifs. We discussed sending yourself the lotto numbers on time, by a day. We discussed it, I realized the paradox of it all. If you decided to conduct this kind of experiment, and then the next day, you send yourself the numbers for lotto, at the time of beginning the experiment, you'd get the numbers before you had sent them back. Then you can go on to and win the lottery, so What happens if you choose not to mail the numbers to yourself? You'd have already won the lottery and so why bother to send the numbers.

The problem is what is the source of the numbers from? If they only came into existence as you considered sending them out, is time travel a thing that's only in our nimbleness and thoughts. If so, could we then not learn to manage time simply by being aware of it?

This is a very similar phenomenon with the double-slit test in the sense that, by either looking at or not looking at the light wave (by either sending or not transmitting messages to the past) by observing the waves they are particles that travel that travel backwards through time, before they even were viewed.

If you choose not to transmit a message, will the message be reversed that of if the message never got through at all? Or , is it the case that time is only a function of the way our brains interpret what we consider to be reality, or is it that reality only exists in our minds? If this is the case, can we not be able to alter the nature of reality by just being aware of it?

The more I study and grow, the more convinced I am that everything happens in our minds. The way we perceive Time as well as our understanding of Reality both reside in our consciousness. My reality might not be identical to yours due to the way that our brains organize it all in accordance with our genetic makeup and personal experiences. I believe that, through studying, learning through research, practice, and assessment the human brain is in the near future capable of understanding the concept of time and space, thereby creating our own realities in the way we like, simply just by contemplating it.

It's been said that the definition of reality is when groups of people have the same experience and can smell, see and feel the same objects.

For example, if could see a forest but there was just me as the sole person who could see it then it's not real . However, when other people can feel it, see it, and even touch it, then it's real.

The evidence from others that something is true is a reality because of faith If you're told that something is real and people who confirm it are true, then it's actually true. Personally I've never believed in that and I still don't, but I can understand how Social Programming carried out by media and governments could make use of this advantage to their advantage.

A little off topic but it's not a big deal. I'll cover certain aspects of this within Social Programming.

So, I'm going conclude that , after analyzing everything I don't think that we are living inside an artificial world, but the argument in support of it raises some interesting questions.

I'm also quite enthusiastic about the future and where it's going thanks to all the advancements in technology and the latest discoveries that are being made.

In terms of reality I believe that we are the ones who create our own choices regarding the reality we live in. Life begins in and finishes with brevity and all that is in between is merely a reflection of the decisions we have made.

Finally, the double slit test is the most intriguing thing I've seen, mostly because it is one step on the other side with regard to the actual evidence.

There's no debate about what it does , but no one is aware of the reasons. It's similar to finding an answer without knowing the query.

Chapter 3: The New Theory Of Everything

In the past 500 years people in the west thought that the earth was flat, and that it was in the middle in the Universe, all stars and the sun revolved around it and that an God was an anthropomorphic God was the one who created it. Then were Christopher Columbus [1451-1506], Ferdinand Magellan [1480-1521], Johannes Kepler [1571-1631], and Galileo Galilei [1564-1642], and the beliefs of these men changed. The earth was then believed that it was round. The sun was believed to be the center of the universe. an God was thought to be an anthropomorphic God was believed to have created it.

Material substance--matter--was all there was, and intelligence and consciousness did not, and could not have existed, until evolution produced a brain. This was the accepted scientific view of the world until the 20th century. The universe was believed to be smaller than it is to be. It was believed to have always existed as well as the sun thought to be the central point of the universe. The scientists believed that they believed that the Milky Way, what we today recognize as one of many galaxies was the only thing in the universe. Furthermore, the most educated people believed that life came through a random accident, and that intelligence and consciousness didn't exist before evolution created the brain.

In the following years, Edwin Hubble [1889-1953], who is the reason why Hubble Space Telescope is named Hubble Space Telescope was named, altered the way reality is perceived through the discovery and report that galaxies are numerous in addition to the Milky Way, that the universe is expanding as well as that our sun not the only one in it. Hubble released his initial work about the relation between distance and red shift in 1929, which revolutionized how we view the universe as well as our role within it. However, many still be of the opinion that consciousness and intelligence were not present until evolution brought about the creation of a brain.

However, our beliefs are being transformed yet again. A significant shift regarding our perception of our universe's origins is happening and I am hoping that this book can inspire at the very least a few brave scientists to investigate the maths, or whatever calculations are essential to justify the new cosmology that this book proposes, i.e., a new theory of the creation of the universe and living sufficient to convince die-hard Scientific Materialists that intelligence and consciousness were present prior to the time that evolution developed the brain and actually are the source of all life and the universe.

The majority of the work is already done. John Samuel Hagelin a Harvard-educated Ph.D. in Quantum Physics has pointed toward this theory by studying the unified field, which is believed by many scientists to be pre-existent prior to the Big Bang, to the realm of pure consciousness, referred to as "Veda" as described in the ancient texts of religion in India known as the Vedas.

Below is a hyperlink to an YouTube clip of talk of Dr. Hagelin.

https://www.youtube.com/watch?v=4u3f7_p1i8c&t=972s

The Dr. Hagelin maintains that the unifying Field and Veda are the same thing. Two paragraphs that are a direct quote from the write that was published under the video:

Two science, one ancient and subjective, and the second contemporary and objective define the manifestation of creation as a reflection of inexhaustible dynamism, encased within the infinity of the fundamental field. Physics defines this relation through the lens of the unified field and the vacuum energy and Vedic Science in terms of Shiva and Shakti.

Similar to that, at every stage of creation in the physical world both descriptions of nature's function match exactly. Physics describes three superfields that give the basis for the five types of spin that define elementary particles, the resonance frequencies of the unifying field and the building elements of creation. Vedic Science speaks of the three Prakritis or Doshas which give birth to five Mahabhutas which form the basis of the universe.

Let me get straight to the point and try to convey in basic, everyday language the theory I believe is in line with and is backed by the above statements. I am saying this because the ancient Rishis from India, sages who lived thousands of years ago believed that Veda that was distilled into the everyday language of consciousness. It is the basis of of all that is. Also the Rishis believed that consciousness was the basis of the universe.

This book was created to present and investigate theories that are in line with Dr. Hagelin's view in that the Rishis' beliefs were right. The theory also supports the

Rishis' beliefs of Veda (consciousness) constitutes the basis of every living thing.

For humans, Veda, or consciousness manifests as the Self, also known as the "I am" as well as the unseen observer in the behind your mind as well as my own mind. But don't get it wrong. the Self (Veda which is also known as conscious) is not a person's mind. Veda is the basis of existence, is what is the guiding force that allows you to open a page in this book. It is the factor within each one of us who examines the thoughts that come up within the mind and decides which thoughts to pursue or discard.

It is difficult to overstate the significance of this realization and the implications are enormous. A few of these will be discussed in future chapters.

What existed prior to the Big Bang?

Does this all fit with the majority of scientists think today?

Scientists have suggested that the idea that led to the creation of life originated in the quantum fluctuations that triggered an explosive chain reaction, the Big Bang that led to the changing universe we live in present day. The theory, however, is based on quantum mechanics. According to British biochemist Rupert Sheldrake said in a now-banned TED Talk, "[Scientists today suggest] we should be given one miracle free of charge and we'll show the rest. The one unpaid miracle that they offer is the creation of all matter and energy in the universe, along with its laws, out of the smallest of things in a single moment."

It is enough to say that instead of explaining the existence of things and the universe, current scientific theories about how the universe began have put things in an extent that begs the concern, "What existed before the beginning?" Could it all be a result of nothing? While that's what most scientists be believing, it doesn't seem to make sense. The song from The Sound of Music goes, "Nothing comes from nothing Nothing ever will."

Instead of starting from the beginning of nothing, it would seem reasonable that the problem of explaining the existence should be focused rather on the definition of a self-existent base of existence without explanation needed. Certain physicists have suggested that the real ground in reality lies within the the quantum dimension of particle which are constantly forming out of existence. While this kind of reality is certainly there however, there is no apparent justification for why the basic reality should be governed by quantum physical laws. An even deeper explanation appears to be necessary and what seems to make sense my mind, as stated in the previous paragraph it is that conscious awareness is the foundation of existence. The way quantum particles spewed out of the sky came to become the basis of the universe calls for an explanation, but the fact is that consciousness is able to explain itself. The unique aspect in consciousness is the fact that it doesn't appear to be grounded in anything other than its own self. Self-producing consciousness is in that it is only within and through it. According to Rene Descartes [1596-1650] famously declared, "I think therefore I am." In the same way there is nothing that must be beyond consciousness to be proven.

Diehard Scientific Materialists will have difficulty renouncing the notion that brains are the source of consciousness, but anyone who is familiar with the findings of researchers at the University of Virginia touched upon in the first chapter of Chapter Two will know that's not the scenario. It's a pity that the findings of this research haven't been widely reported--I believe this is that's because science journalists are scared of being slammed by Materialists. But as I've mentioned previously it is the brain that is a receiver which integrates consciousness with the body which could be called"the "mind-body complex" as it's an apparatus or vehicle which allows your consciousness to exist and function within the physical (physical) space. If you're not convinced visit YouTube and type in the following word into the search box: "Dr Bruce Greyson Consciousness independent of the brain." A talk by the Dr. Greyson that goes into depth about the research of UVa should be at or near the top of the page. It is Dr. Greyson, by the fact, is not a New Age Looney Tune. The Professor is Emeritus of Psychology as well as Neurobehavioral Sciences in the University of Virginia School of Medicine.

All Is One

If the brain doesn't generate consciousness, from what do you think it originates from? In the previous paragraph consciousness is the Source that is the Veda or source of being that creates and informs the reality. We will look at evidence to support this. It is said that the British quantum physicist Sir James Jeans [1877-1946], wrote: "The universe begins to appear more like a great thought than a magnificent machine." The scientist was not only pointing to something, he struck the nail right on the head. He was aware of the quantum physicists all know that there isn't any kind of matter in itself, no separate entities that are made up of solid material. All that is there are energy and vibrations simple and pure. Since physical reality is based on vibrations, there is no separation between them. Everything is interconnected. There is no boundary or a line that stops one vibration and another begins. Mystics have said since the beginning in the dawn of history, "All is one."

Furthermore, if consciousness is fundamental, then consciousness is the source of everything, including the fundamental elements of time, space mass, charge and space. Like Sir James Jeans seems to have mentioned in his earlier comment that the universe looks like an idea and the physical world we live in could be like an idea or a dream the Source is experiencing.

Take a look. When we sleep Our minds make the dreams we have, with their images and all the trappings they carry however, they are to be real. If a car you vision is poised to hit you over, say you're sure it will hurt and that's the reason why you awake. But only when your eyes open do you realize that it was just a fantasy.

Think about the following. There must be a person in your dream who's perspective what's happening is seen. A dream can't exist without someone watching it and in your dreams this character is you. Therefore, the universe and the world are the result of a "dream" within the mind of God, who later in the future I will call "the Great dreamer." In your dreams and mine, there must be a person in the dream to be seen. Who is this character?

You are that character and I am that character.

Humans, like all living things, are creatures in the Creator's dreams who observe the world around us. Because everything is connected and one with the Creator as well as us, we are The Creator's eyes as well as ears. We are the vehicle through which the Source understands his/her creation.

The reason this might be difficult to comprehend is because in our culture we think of ourselves as distinct entities. We see our universe in terms of "out out there." We see this book as well as that Kindle device or the phone you're holding as an individual object. However, as we've said before quantum mechanics suggests this isn't how things actually are. We're united, we are all part of the same dream. The idea will begin to become apparent and begin to be more clear when you don't dismiss it without thinking and really contemplate it in a rational way.

Panpsychism isn't on the mark

Let me emphasize that the idea presented in this book is totally different from the theory dubbed "panpsychism," which some Scientific Materialists are promulgating in attempts to explain how matter can create consciousness. Panpsychism is the concept that is discussed in Chapter One which was presented in the 2013 Ted Talk by David John Chalmers. It's the belief that everything that is material even the smallest of objects, has an element of consciousness. At the heart, the quark or atom photon or electron is a small amount of consciousness.

The theory suggests that consciousness is connected to the processing of information. The more data being processed, the greater the level of consciousness. This would be the reason the human mind is more advanced consciousness level than an ear of corn or a mouse. of corn. Based on this theory sophisticated information processing activities will in turn produce an advanced consciousness. This is in fact consistent with Bostrom's theory since a computer that processes massive amounts of data would also be aware.

Based on the findings of research and observations made from DOPS scientists at the University of Virginia School of Medicine however, the panpsychism theory isn't true. In order to summarize what UVa researchers have discovered that it is the brain that receives for consciousness, but it doesn't make it. What else can children do to remember their previous lives, when in between, they were without a brain? How can people be aware of what nurses and doctors are discussing and also be aware of what's happening in the operating rooms, in the event that their brains are flat and are clinically dead? If the brain has been flat-lined, it can't be processing information.

Consciousness Only Is Conscious

It is equally important to realize that, in their quest to solve the mystery of creation Materialists have approached the subject from the wrong angle. Physical reality is created and is contained in consciousness and not the reverse. Matter doesn't create consciousness. Consciousness creates matter and that is why consciousness is the best place to start an investigation.

To understand this idea, it could be helpful to recognize that it's impossible for any person or thing to directly experience physical reality this is a sign that physical reality doesn't exist and cannot exist beyond consciousness. The way we experience this dimension is through the senses of hearing, sight scent, taste and sight -- the five senses of the physical body-mind complex. There is no physical object without using some of these senses. In addition our consciousness and the Veda, the Source, are one and the same, the "I AM" inside each of us.

Only one consciousness is present that is the Source. because it is the only thing there is, it can't be seen outside of it. The Source, also known as Infinite Mind and intelligence, has discovered an avenue. We are the vehicle through which the Source creates itself.

Chapter 4: The Implications Of New Theory

Knowing it is that the consciousness of a person is basis of our being and that the actual Self within each of us is the ground of our being, is a blessing to the world once it is well-known and widely accepted. Furthermore, it could bring joy for you and people who don't hesitate to embrace it. The obvious consequence of the fact that every mind is in sync with the source is that everyone is at a fundamentally equal level, and that our beliefs and thoughts determine our individual experiences. Skin color, nationality and other cultural background are not a reason to limit anyone's ability to participate. In a democratic society the only ones who suffer are those who believe themselves as victims. Knowing that your mind isn't confined in your skull can help for you and others to accept and accept information that can open doors you might not have thought existed. Take a moment to think about it. The Source is the Source creating your reality. Your mind and that of the Source are interconnected and affect each other. You are in control and you can follow the stream created by the Source or oppose it. It's yours to make. The majority of people alive today are unaware of this as it isn't part of the perspective they believe in. I hope this book will aid those people who have read the book to see the world through their eyes.

To demonstrate the ways our mental frameworks could make us blind, take a look at the things Charles Darwin found on his journey to Micronesia during his journey aboard the Beagle. In the time of Darwin the inhabitants of these islands were so far to the outside world, they had not had a glimpse of a ship. Darwin and the other members of the Beagle landed in dinghies. Natives were not having any difficulty being able to see the dinghies. They, after all, had small vessels. But they couldn't appear to be able to be able to see the Beagle docked off the coast, even when they pointed it out. A boat of this size was not a part of their mental picture. This meant that it was unnoticed by them. It's the same in the present, for instance with regards to ageing and illness. As it is today, the way people view the world keeps it from being revealed. The people who read what's presented here may not "see" the issues I'm speaking about. All I want is to have you put aside doubts while reading.

The spring of 2000, a shocking realization occurred to me after an interview on local radio regarding one of my novels. It was a night. I was exhausted after having spent an hour trying to be entertaining and funny. When I returned to home, I stopped by the nearby Seven Eleven for a bottle of beer. A sign caught my attention as I approached the cash register.

"We I.D. younger than 27 years old."

I was able to get my spot in the line, amidst two teenagers who were drinking Slurpies. A friend from college stood in my place and we exchanged pleasantries.

When my turn came, I placed the bottle on the counter, and then went to my wallet.

The clerk sat me down. "Sorry I'll need to check the I.D.," she told me.

"Excuse you?" I said.

"I'm going to have to verify the I.D.," she continued.

"You're kidding," I said.

The woman let loose an annoyed sound of sigh. "No I must check your I.D. before I'm able to sell you the beer."

I put my driver's license in it in her hands I turned my attention to my friend and gave her a slight shrug. Her mouth gaped. "It's real," she said, shaking her head. "You truly look young."

When I got home, I drank while watching at the police and contemplated the fact I was asked to prove I was old enough to purchase alcohol. The truth is, I was fifty-five at the time. That's more than double what clerks were required to I.D. It's true that I felt younger. Even now, over two decades later, observe hardly any difference in the way I feel today as compared to how I felt when I was actually twenty-seven years old.

Following that experience I began to think about what was it that made me look young. After some time, the possibility came to mind. Thirty years earlier the age of 25, when I was I had read an article on a study that looked at people who had been taking large amounts of vitamin E for about ten years. The study stated there were no signs of aging were evident within these people. Therefore, I went out for a purchase of a bottle and have taken it ever since.

For a long time I believed I'd never become old. In fact, for a long time it appeared that I was young.

Then, I discovered that scientists had concluded that Vitamin E as a pill can't be shown to reduce aging. Like is often the case, the latest studies contradict older ones. However, I continued to believe the information regardless.

According to recent reports that have appeared in the media, we've almost come full circle. Researchers aren't ready to claim that vitamin E can stop aging completely however, new research suggests that the consumption of vitamin E has a positive effect on the incidence of cancer and heart disease and can help reduce many health issues. But I've come to believe that it could have been beneficial for me in large part because of its placebo effects. However, it did work. Thirty years earlier, I read an article saying I wouldn't be old if tried the supplement. I expected it to do the job and it did. If I'd read an article that claimed the anti-aging benefits from vitamin E weren't just a flimsy claim I'm not sure I would have had the same results.

The power of belief is extremely potent. The efficacy in the use of placebos was proven repeatedly in double-blind scientific tests. The placebo effect, the phenomenon of people experiencing better health after taking dud pills -- is observed throughout the world of medicine. A study claims that, after a myriad of studies and thousands of prescribed medications, and the tens of billions in sales sugar pills are just as effective in fighting depression just as the antidepressants, such as Prozac, Paxil and Zoloft. In addition, placebos trigger radical changes to the same regions of the brain that are affected by these drugs as per this study. For those who be unsure this research proves beyond doubt that our thoughts and beliefs are able to cause physical changes to our bodies.

Additionally, the same study shows that placebos frequently surpass the drugs they're competing against. For instance during a clinical trial in April of 2002 in which the herbaceous treatment St. John's wort to Zoloft, St. John's wort completely cured 24 percent of patients suffering from depression that took it. Zoloft was able to cure 25 percent of the patients. The placebo, however, fully healed 32 percent.

What one believes to be genuine medicine creates the expectation of results and what people expect to happen typically happens. This book will discuss the reasons. It's been established by example that in the cultures that have a belief in voodoo or magical practices individuals will die when they are cursed by the Shaman. This kind of curse isn't likely to have any effect on a person who isn't convinced. Faith is a powerful thing. It is the primary ingredient to manifesting your dreams. An experiment conducted on the Discovery TV Channel, for instance, offers a clue. In this instance two researchers conducted the exact ESP experiment within the same laboratory, with the identical equipment. They took great care to ensure that everything was identical, apart from one difference. One researcher believed that ESP was legitimate, while the other didn't. Both tests were overseen by independent observers, including those from the Discovery Channel TV crew.

The test that was conducted by a person who was convinced ESP was statistically significant in the amount of correct scores, which suggests that the study was successful. The reliability of ESP was proven scientifically. However, the right hits during the test with the skeptical Thomas researcher fell within parameters which could be explained due to chance, which is why the experiment was not able to prove the legitimacy of ESP. Apparently, the one and only variable--belief--made the difference. One researcher was convinced in the theory, and the second did not. Both got the results that he had hoped for.

The same is true for praying by Christians. Prayer works. Prayer releases thought in the mind. Prayers can give the spirit, or"the Life Force," an additional enthusiasm that helps to increase its natural ability to organize things in a manner that is beneficial to the living. Soon it will be clear to you precisely how this works, and in an upcoming chapter we will cover in some depth the effects of prayer as demonstrated in scientifically-constructed, double-blind experiments.

We do create our reality. This is explained in the lectures I was able to find by a person named Thomas Troward. He delivered them first on the Queens Gate at Edinburgh University in Scotland in 1904. They were dubbed The Edinburgh Lectures on Mental Science They offer a concise rational explanation which is in perfectly with the results of research regarding prayer. They show that the distance between those praying and those whom they pray for isn't an issue, and the person who is who is being prayed for doesn't need to be aware of the prayers made on his behalf. The way prayer works is straightforward however, let me begin by laying the foundation before I present it to you.

It's helpful to begin by examining the distinction that exists between what we consider to be "dead" matter, and something that we consider to be alive. Plants, like one that is a sunflower, possess an attribute that makes it different from steel. The sunflower can turn towards the sun with its own energy. After the first time it is picked it is characterized by a type of glowing. This characteristic could be referred to as the Life Force or Spirit. However the steel appears completely inert. But at the quantum level, it is alive with movement. Quantum scientists have revealed that energy or motion is the basis of the entirety of matter. The molecules and the atoms are not tangible objects. These are both energy. Vibrations. Many would argue that the entire world is alive as if it were one giant thought--the idea of an infinitely large mind that organizes intelligence.

However, based on external appearances, the sunflower appears to be alive, but the steel isn't. There is no way to argue with this. However, one could say that a plant's level of "aliveness" differs than that of an animal. Think about the difference in the quality of life between a sunflower earthworm, or the goldfish. Each is progressively more active.

Let's now include a dog, a three-year-old child and a stand-up comedian who is on the late night talk shows. Each one has a higher degree of intelligence. Therefore, in a certain way the level of "aliveness" can be determined by the degree of intelligence or awareness displayed, in terms of the power of thinking.

As mentioned earlier the concept of consciousness, intelligence and thought, also known as mind is the foundation of and produces the whole universe. It becomes more apparent to us--we perceive it better--as consciousness is more aware of itself. The distinctive characteristic of the spirit, or life is thought. And the characteristic nature of matter, like in the steel piece is its form.

Think about for a moment the difference between the distinction between thought and form. Form is the term used to describe the occupation of space as well as the restriction within certain limits. Thinking (or life) does not imply either. If we consider thoughts or life in a particular way, we connect it with the notion of space being occupied and therefore the elephant can be considered to comprise a greater amount of living matter than mice. However, if we view existence as the reality that it is "aliveness," or animating spirit, we don't consider it to be being in space. The mouse is just equally alive as the elephant, despite the differences in dimensions. This is an important fact. If we could imagine something as not being in space, or having an undefined form, that thing has to exist in its entirety everywhere and everywhere, that is everywhere in space at once.

Thought and life do not take up space, but also transcends time. The definition of time as a scientific concept is the duration of time occupied by an individual as they move between one place across space. In other words, when there is no space, there is no time. If life or thought isn't devoid of space, it also has to be free of time. In the end, all thought, or life exists everywhere simultaneously in a universal now and eternal now as the scientists who have studied this and solved it mathematically be able to agree.

How can this aid us in understanding the ways we make our own reality, as well as what prayer is?

In the first place, it is implied in the previous discussion the fact that we have two types of thoughts. They could be referred to as lower and higher or subjective and objective , since the thing that differentiates the higher and lower is the self-recognition. The worm, the plant and maybe goldfish have only the lower type. They do not know who they are. Maybe the dog, and definitely the comedian and the boy are both. The more advanced level of self-awareness can be found in increasing quantities similar to ascending a scale.

The second mode of thinking The subjective mode of thought is the subconscious or mind everywhere that is, among other things, regulates and supports the mechanisms of life in all species and every single person. It triggers the plant to expand towards the sun, and then grow roots that are pushed into the earth. It triggers the heart to beat, and lungs to breathe in air. It regulates all the involuntary functions that are a part in the body. As we'll discover, it is in charge of many more.

The fact that this type of thinking is prevalent and at the same time, is in accordance with the theories that was proposed by Carl Jung who maintained that humans are part of a common mind. Additionally, we all have the capacity to have our own individual subconscious mind , which merges into a collective mind, which contains archetypes and other archetypes and then blends with the ground-of-being, subjective mind mentioned above. Minds that are conscious of themselves serve as the ones that create the kinds of our thoughts, which make us aware of ourselves. The different kinds of minds are inextricably connected, because they are born from the subjective mind, which is non-dual. Non-dual and transcendent the mind of this type does not differentiate between evil and good. It was the way it was before humans reached the point where they were able to take a step back and contemplate the distinction between what is "good" and the things that weren't. This point in the evolution process is described literally in the story of Adam and Eve when the couple consumed the fruits from the tree of understanding of evil and good and realized that they were naked. In Shakespeare's Hamlet declared, "Why, then, it is not yours since there isn't anything in between good and bad however, thinking can make it the case." It is the gradual rise of self-awareness from this non-dual mind is evident in our analysis of the earthworm, plant goldfish, dog comedian, boy and so on down the scale.

Let us now consider the main point that was presented by the lecturers. The conscious mind is able to exert control over subjective thoughts which is the source of our perception of reality. I realized the truth of this in the first semester of college, when I was able to manipulate other people. I would place a student in a trance, and make him believe the he was a chicken an animal. Much to the delight of my students I would have him respond in the same way.

Hypnotism works because the person who hypnotizes is able to bypass the conscious mind of his client and directly addresses the subject's subconscious mind. The mind of the subject has no choice but to accept as the real world what is presented directly to it as factual by the conscious mind. Since it is completely subjective, it can't take a step back and look objectively at things. It can only be deductive reasoning. This is the kind of thinking that proceeds from a source (the conscious mind's instructions) towards its final goal, with the mind of the golden retriever. It is not able to stop to think or consider. This is the reason criminals might employ when the commission of an offense. A criminal might be in a room and observe a man weighing his money, and then think: "I need money, so I'll steal his. As the man is trying to protect the money, I'll remove him. He'll be shot. He'll fall to the ground. I'll take the money and walk away. I'll walk out by my window." The right and wrong good, bad and right aren't discussed, just how to reach the desired outcome.

However the conscious mind being self-aware and objective, can be able to transcend. It is able to reason inductively and deductively. Inductive reasoning means to go back from the result to cause. A detective from the police force, for instance, will arrive at the scene and start thinking backwards to try to figure out which crime was committed and who could have committed it.

This means it is that subjective thoughts are controlled by that conscious (objective) brain. With the utmost integrity the subjective mind works tirelessly to reinforce or transform into reality what it believes is real. Because the mind of the individual merges with the mind of the ground and is everywhere and everywhere, it can influence events and situations to ensure that what it believes as true, will be true. Thus, for instance when I believe that I'm sickly and I believe that I am sick. In the event that I think being in the draft, I will get cold, then I'll catch colds as I sit in the draft. However, if I believe I am rich and that I am entitled to be wealthy because of my birthright, I will be rich , even if it is not the scenario. If I believe that I are not lucky, then I'm not lucky.

This also clarifies how, what and why prayer is effective. If people who truly pray believe that their prayers will have an effect on the world the prayers will. Their belief is reflected in their subjective minds. Their personal minds are merged with the mind of the ground. The more people pray or believing the more the impact. The mind of the individual for whom they pray is also a part of the mind's ground and this mind goes to work in order to produce positive outcomes.

It doesn't make a difference if the person praying is sitting at a bedside or in a different location. As mentioned above thoughts, and hence prayer is everywhere at the same time. It isn't local. This is the reason why prayer isn't restricted by distance.

A majority of people get and are lulled into believing they have no control over their situation. However, they are the ones who create their own circumstances through their beliefs and thoughts. This is the message from the Edinburgh Lectures is simple. Change your thinking and your situation will change. Also, while you're in the process A few prayers with a good intention aren't going to hurt.

Recalling my own experiences with the power of faith along with aging as days passed after the episode from Seven Eleven, something else came to my mind. Beginning a new chapter in life could be the result of subtracting years from one's chronological age. In 1993, I began to realize that I had become stagnant in my work. I started to have a regular visual of myself coming back to the same track over and over. You could call it the feeling of deja vu. I had a successful career and was the president of my own advertising agency. I was earning the highest salary of six figures and was featured within Who's Who in the Media and Communications. I'd achieved what society and our education system suggest is the main objective of life, and the only path to happiness and fulfillment. I'd picked a field of study and made it to the top of the heap.

And , as many who breathe the rarest air discovered that it wasn't what it was made out to be.

Don't get me wrong. I love the process of creating and being creative is what advertising about. When you're successful, which is the case in many fields of work, at some point, you may stop doing the things that made you successful. You become a supervisor for others who are having the pleasure, but you also have the stress. So , I sold my advertising agency and went on a break for a year before I started doing some work in marketing communications due to the fact that, by then I was aware that I was in need of money to provide for my family, put the roof over our heads and pay my tuition. However, I organized my marketing communications and advertising tasks so that I was in a position to focus on the aspect I enjoy and that is the creation of advertisements and campaigns, and at the same time, I could do what I really wanted. It was back in 1993, I felt I was being called by something.

I've found that if you desire something and really would like it and stay focused to the possibility that it will present itself. In the late Joseph Campbell [1904-1987] labeled this chance "The calling to go on an adventure." This opportunity is likely to come regardless of whether the desire is recognized by you at a conscious level or if it's buried in the subconscious mind or your soul. Then you'll have a decision. You may follow the path and benefit from it. Or , you could decline the invitation and in that case, you'll stagnate and eventually pass away, perhaps even physically. This is a sign of many health issues that's not visible to us. Here's a key fact that if you take the challenge of adventure is to take the life you want over old life and dying.

Myths from all cultures tell the same story over and over every time in the form of a different cultural interpretation. This is not surprising as the urge to explore is something we all encounters, often multiple times throughout our lives. We are often compelled to leave the security as well as security that is our base, which is what Campbell described as the hero's "Ordinary world," and venture forward into the unknown, where monsters, dragons or demons of a kind or another have to be dealt with and defeated. With the help of supernatural or unseen powers, the hero that is determined to go forward will always succeed, only to return to his or her home higher-leveled than when they left and with the benefit of a whole new degree of awareness. The purpose of life is growth and that is the reason we leave our home, the spiritual realm of spirit, to come to the rough and tumble world known as Earth. It's to experience both the good and bad as well as to confront difficulties and overcome them as that's the process that allows us to grow and develop. The majority of our society does not realize this, despite the fact that it is a constant reminder that it is the right time to be on a new degree of knowledge. The denial of this essential element of our lives is especially tragic because the worst consequences are always a result of our unwillingness to acknowledge the call.

Don't trust me to tell you this. The warnings are in the myths of all ages. Refusal transforms what could be constructive and positive, into negative. The person who would be hero-like loses the ability to act and is instead a victimized victim of boredom, hard work or even a prison sentence. King Minos did not accept the invitation to sacrifice his bull for instance, as it would have symbolized his surrender to God. Naturally, the king did not realize that this could result in his being elevated to a higher level. As an modern-day professional or business executive, the man was trapped by the conventional wisdom and tried to get out of the way with dedication and hard work. In fact, he was successful in building a home for himself, in the same way that many professional executives and managers are now building mansions in suburbs. However, it ended up being an uninhabitable place, a home of the dead, a labyrinth where one could hide and escape from the dreadful Minotaur.

Take a look at what transpired to Daphne the gorgeous girl who was pursued by the beautiful Greek god Apollo. He only wanted to become her lover and he pleaded with Daphne, "I who pursue you are not your enemy. You do not know from whom you are fleeing. It's because of this that you flee." The only thing Daphne required was to acknowledge the invitation, and then the most beautiful and abundant love was hers. Shetoo could have been in an encounter with the Divine. But , as you may have guessed she didn't give in to. She continued to run and eventually changed into a laurel plant which ended her.

Two stories pop into my head from popular culture, which provide an example of what I'm hoping to convey in response to the question to learn from experience. The first story is The Wizard of Oz. In this film, Kansas is the heroine's (Dorothy's) "ordinary world," where we find her wishing for the possibility of a better life that she believes could be "somewhere across"the rainbow."

There's Miss Gulch who wants to get her pet, Toto be destroyed as Toto has attacked her. Miss Gulch steals Toto away to accomplish this however, the dog escapes and returns home. Dorothy goes off along with Toto and eventually returns to her home only to be removed from the security and safety of the farm on which they live by the twister. Dorothy is transported to a foreign, secluded location, called in the Land of Oz, where she finds friends, overcomes and overcomes challenges and gets into the ocean's belly as the castle of the witch. After that, when everything seems impossible, Dorothy succeeds in what she was destined to accomplish--she catches her witch's broom. Then, even if the wizard proves to be an obfuscator this creates what storytellers refer to as "the dark moment,"" Dorothy is able to return to her home.

The story comes to an end the reader discovers that Dorothy is evolving because of the adventures and experiences she has had and has a profound understanding that she didn't have before. Dorothy is now aware that there's "no home as home."

The second story, in my opinion, is an allegory of what the life of Earth is all about. I'm referring to the film, Groundhog Day.

In the beginning of the tale the protagonist actor Bill Murray's character along with a television crew, have left Pittsburgh and their everyday lives and traveled towards Punxsutawney, Pennsylvania to cover the annual Groundhog Day ceremony held there every February 2nd. As the happenings of the day take place, the character of Murray shows his self-centered character who is a jerk.

After an arduous day Murray along with the crew of TV are stranded in Punxsutawney by a storm and are forced to stay the night there.

The next day, when Murray's character awakes the next morning, he discovers that it's not February 3rd but rather February 2nd, all over again. Everything is exactly the same as it did in the scenario that Murray's character sees as the previous day, however all other characters sense that it's occurring for the first time.

Murray is reacting in the exact same way, self-centered and egocentric and is then confronted with the fact that he's caught in a continuous loop, as Groundhog Day repeats time after time in what appears to be an inexplicably perpetual loop. As time passes Murray's reaction to events shift. Over what appears to be like an interminable amount during Groundhog Days, his actions become more streamlined to the point when he is able to handle the things in a loving and professional way. In the morning, after having completed all of the tasks in what appears to be the most appropriate way, he awoke to discover that it's the 3rd of February.

He's now able go on.

I view this as an allegory for the life we live on Earth since I believe we live in the same place time and time again and are faced with similar problems and obstacles until we are able to handle them in the most effective way that are possible.

Let me share one more story. In this story the story, a young man is given an invitation to go on adventure, but he rejects the offer. It is a story about Jesus and is included in all three gospels synoptic. The story is drawn from Mark 10:17-23. It is the New International Version (NIV) translation:

When Jesus began to walk when he left, a man came towards him and fell to his knees in front of Jesus. "Good instructor," he asked, "what should I do in order to be heir to Eternal Life?"

"Why do you consider me"good"?" Jesus answered. "No one is perfect, except for God only. You are familiar with the commandments"Do not commit murder and do not commit adultery, don't steal, don't provide false testimony, don't fraud, and honor your parents and your father.'"

"Teacher," he declared, "all these I have kept since I was a child."

Jesus was looking at him and adored him. "One thing you're missing," he said. "Go to get rid of everything you own and give it to the needy and you'll be blessed in heaven. Follow me, then come."

When this happened, the man's smile dropped. He left in sadness, since he owned immense wealth.

Jesus was looking at the world and told His disciples "How greuly it's for those who are rich to gain entry into in the kingdom of God!"

The disciples were shocked by the words of Jesus. However, Jesus repeated his words, "Children, how hard to be a part of God's Kingdom! God! It's more simple for a camel pass through the needle's eye than for a man of wealth to gain entry into in the kingdom of God."

There you are. If anyone ever was contacted to go on an adventure and adventure, it was this wealthy man. If he accepts the call and accepts the invitation, he will be heading towards Eternal Life, which I consider to be a union of The

Infinite Mind and what Jesus also described as getting into The Kingdom that is Heaven. Also, as in the cases of Minos and Daphne the guarantee is a young man is able to answer the request, he will be able to develop in time and experience joy of a connection with the Divine. However, first like for both the two, he has to give up the material things that he treasures and set aside his ego in the same way as Bill Murray's character was able to do when he appeared in Groundhog Day. But like for Minos and Daphne as well as the majority of us, the character was overly attached to his personal wealth and self-centered confidence to take this step.

As I reminisced about my own personal story, I was, exhausted of the ads and eager to ascend up to a new level but shackled by gold handcuffs, just like the young, wealthy man in the above story. I was ready for the opportunity and, naturally, it did come. It was difficult to walk away from this precious treasure and yet I did. I began writing. And I was hooked.

Like all hero's adventures It was a bit scary to take the first step, to respond the request, and the journey was even more terrifying as it progressed. Then, it wasn't long before there was a lot more money was flowing out than it was coming in. I began working as a freelancer to slow the flow however, I still had to tap into my savings massively before things started to level out. The process I had to go through was not enjoyable however, let it be enough to say that I had to face my own demons and dragons and confront my anxieties that made me think I should put my nose back against the concrete of everyday life and find a job. However, just like in every hero's story when things got really, really difficult and unimaginable hands, the assistance of God came in. But , this story you know, dear readers, will be waiting for a future book.

I'm sure that, like other people who have gone on the same journey, I've returned to the starting point. True I, just like the great heroes of old I'm now on a greater level of understanding, both spiritually and intellectually however, I am it is sad to say that I'm not more prosperous financially. However, I am more wealthy in several ways than before I accepted the invitation and I am certain that, as long as pursue the life I chose to lead, the invisible hands will be able to clear a path for me, and ensure that I'll have enough food to have food and shelter above my head. I've stepped into the Kingdom, and am now a faithful subject and am reaping the rewards. I'm no longer fearful of death or poverty. Instead of doubt and fear I feel the power of God that is ready to assist me when the going gets difficult.

Chapter 5: The Social Programming

Are we living in the computer system and are being controlled by the social Programming that is carried out by governments via the media?

As I mentioned earlier, I've been online for hours watching videos and reading about Social Programming by our governments and yet another event leads to another creating a more complete picture.

It is obvious at the start this Social Programming is very real and has been extensively researched and utilized in a variety of aspects of our lives , without any prior knowledge. It has become until today a massively influential role in our society.

But, despite that it seems to be harmless , and perhaps somewhat beneficial, so it's just a few minor incidents in our lives that can help guide us toward a better direction.

until you go a bit further, that's.

The questions I've been trying to answer are, firstly, what are Social Programming, if it's true, what exactly are we instructed to do and who is in charge of what we do, and who profited from it whether it's large government or corporations, or is it just the natural evolution of a society's need to conform to expand, providing an outlet to conspiracy theorists who believe that everyone is under some kind that controls our minds.

If this is a plot does it have any justification for it. Are we being given false information from birth to death, consciously to keep us within the line to provide power to the elite? If so, how deep does this go? And lastly, how does it connect to Simulation Theory and what can we do to stop it.

I'm planning to solve all those questions by writing down the information I've discovered while researching as a large portion of what I've seen and read isn't proved or authentic and could possibly be the work of someone in some kind of conspiracy. I will only write about what

I've observed as well as read which I consider to be the truth according to my understanding and then in the final paragraph I'll provide my own opinions and conclusions based on my personal simple perspective.

What does it mean to Social Programming?

The most straightforward explanation I could find was an experiment that involved five monkeys in cages.

In the cage were five monkeys as well as an elevated stepladder. On at the very top was a plethora of bananas.

If one of them climbed up the ladder to collect the bananas, the others were soaked by cold water from the pipe. The process was repeated many times until the monkeys were exhausted from being sprayed that none them wanted to climb the ladder again.

Then, at the moment, they opened their cages and exchanged one the monkeys for another monkey. Within a few minutes the new monkey began to climb up the ladder. However, due to the cold water , the other four monkeys jumped over the ladder and beat their new monkey.

However, this time no monkeys were able to climb the ladder, and they again went into the cage and exchanged one of the monkeys from the beginning to a new one. Then, after some time after a few minutes, the new monkey began climbing to climb the ladder. The four other monkeys pulled on the ladder , beating the new monkey. The process continued until the initial five monkeys were replaced with new ones. Five monkeys that were now inside the cage had ever been bathed in the cold water, but they would not go up the ladder. Whenever they were introduced to a new monkey and attempted to climb the ladder the other monkeys would fling themselves off the ladder before beating it.

If we could inquire about the reasons why monkeys did this, they would have any idea! Apart from the fact that's how we live our lives.

They didn't know of the freezing cold, therefore they didn't know the reason they couldn't climb the ladder. Consequently, they didn't know the reason behind beating any monkey who tried. It was just an act that they carried on, passing from one monkey group to the next in a matter of minutes.

It could be us in the cage, passing through generations with no one really knowing what drives us to live our daily routines in the way we do it's the way we operate. We're so used to the way we conduct our lives, that we aren't able to ask questions.

The first day of school begins with learning the alphabet , understanding words, as well as the basics of math and numbers. As we progress through our lives, we progress into biology, science or history, religion, and many other subjects, believing that everything we're told is the truth, as what we're also taught from the start is that we are not supposed to question authority, and not to voice our opinions, instead of fitting in as a quiet, respectful and use the bathroom only at times that are permitted, take breaks when permitted and have dinner when it is allowed. You work from every day from Monday to Friday, work for an employer, and when you reach the point of retirement, you can leave school then you can find an occupation, get cars, buy an apartment, get married and have children. It's all created to make money, and then spend it on your expenditures that are always on par with or even higher than your income. The government will tax your earnings and taxes everything else you spend your money on. They it wants you to continue spending and earn more, but not making enough so that you begin to save.

Companies conduct research on our behavior patterns to see if they can make small modifications that will lead us to investing more. For instance, the London's Heathrow Airport has tens of thousands of passengers traveling through the airport every day , yet it intentionally seven hundred seats. If you'd like to settle down, but you can't locate a vacant seat, there are plenty of bars and eateries where you can also be able to spend money. the airport boasts that they make five hundred million dollars a year from this alone.

Supermarkets place the essentials of life such as milk and bread (never in a row) in the back of their shops so that you must go through the rest of the store in the hopes there is something that will draw your eye and make you spend more. Casinos design their layout as an intricate maze that makes it difficult to figure your way out, and they have no windows that let in light which means that you easily lose sight of time, hoping that you'll be there longer and pay more.

If businesses are doing this to boost their revenue so why should us think that governments wouldn't.

On the surface, there are hundreds of tiny things that we see every each day, all put in place by the government. These were specially designed to have the goal of controlling the way we behave. They are often not noticed and are mostly for our advantage, but knowing they exist tells us that the government is active in implementing control strategies and offers us the opportunity to start from a place of control.

The things I'm referring to are the thousands of fake CCTV cameras that are designed to make us believe that we're being monitored to make us more likely to adhere to these regulations (the rules they establish) even though there are millions of CCTV cameras that are genuine and according to some estimates, everyone who is going through our day-to-day routines is captured on camera an average of 70 times per every day.

Additionally, there are things that are very easy to do like bridges with uneven surfaces so nobody can sleep rough and pink lighting in places where they want you to be relaxed and blue lights in areas in areas where they want your to appear more vigilant. One of my favorites is the buttons at the road crossings. The majority of them are fake and perform absolutely nothing.

The lights are completely controlled and the buttons are specifically designed to create an illusion of control.
"Control" is the main word!
Larger online companies such as Facebook have admitted openly the fact that Computer Programming has been deliberately designed to boost dopamine levels within your brain, as dopamine can be addictive, and it keeps users wanting more.
Facebook alone has 2 billion monthly users that are targeted individually with a custom ads.

They have publicly discussed tests they've done in collaboration with some 689,000 users with no knowledge. They have deliberately chosen advertisements and other content to display on their screens which could alter their mood.

These types of companies that earn billions such as Facebook, Twitter and Google are not new in Social Programming, where governments have been trying out this for years.

However, as mentioned there are strong connections between these firms to the NSA and CIA that are both government controlled agencies.

It's been reported that when inventors come up with the idea of a novel invention or idea and apply for patent rights for their notion or concept, there's corruption at the patent office and inventions that could be detrimental to the masses are alerted to the government assist and decide for themselves, and then to throw money at them in the millions to get businesses in operation, then put an unnamed person in place to manage it, and then let shareholders in, then relax and take over the business in the shadows.

I'm not sure if this is true or not, but I have seen a few videos with the same claim, however when I tried digging through it to find the truth, I hit a brick wall. It was almost as if videos and articles were intentionally removed from the web and that's why I'm more likely to believe it's real.

The world's governments are everywhere and are involved in everything and do not leave any stone unturned, consequently it's extremely difficult to determine the degree of social programming because it's not just one thing; it's a multitude of levels that span hundreds of different aspects, with the most prominent being media through your TV, computers, and mobile phone screens.

However, every aspect, however large or small, they all have the same common thread that is to control you, to maintain your docility and guiding you through your the day to ensure that you fit into their plans.

Displaying only the other side of an argument before it's time for you to decide or simply showing the horrors of another country when they need justification for war is only two ways how they can make use of the media.

Numerous studies of the effects of watching television have revealed that what we see and hear on TV is a part of our minds for the remainder of our lives. It is a an integral part in our process of making decisions.

Every time we make any kind of decision, our brains scan through our collective information. One component of the collective knowledge is everything you've ever seen on television. It was discovered that children as young as two years old watch television on average for at least thirty-four hours a week, and that figure only grows as they grow older.

The study also looked into the brainwave patterns while watching TV as opposed to reading books. It was observed that after just a minute of watching TV , your brainwaves switched from Beta and then Alpha waves. Beta waves are linked to the flow of thoughts, logic, and activity Alpha waves are generally associated with dreams and learning passively, this is the time when the right part of your brain is activated and your critical thinking abilities are deficient.

When you stop watching television and begin reading books Your brainwaves immediately return to beta waves. The TV's presence effectively dulls the left part that your brain. Therefore, the person who decides what content is displayed on the television also controls the person who is watching it.

My conclusion from this is that the government does censor the things we learn at the age of a child, and also the content we watch on TV and in the web, and even the content we read in the media. To gain power, money and position.

Governments strive to keep those who are poor, poor and the rich, wealthy but they aren't able to tell us that they're doing this since it could upset the balance. We are instead told that the government is to serve the public and are in the best interest of the people. To which we have the right to choose whom we'd like to be our leader. However, the whole system is a clever way to give us the appearance of control when it's all about keeping the wealthy, wealthy and the and the middle class.

The current, elected government at any time serves the elite, those who set the stage for them to become the bosses at all. those who are among the most wealthy of the world are those who control all the cards. They are the ones in charge of all things, their motives isn't the financial aspect. It's their power to control and control money that is their weapon.

The only solution to stop the cycle is to educate yourself and self-awareness. For instance, if you were told only that there are only two alphabetic letters (A A and B) and presented with the option of choosing one of them, you'd not have any idea of any other option which is why you'd choose one of A or B. It's difficult to make a choice in life if you're never aware of the possibility of a choice which is why learning more than the information you are provided with can open your eyes to new possibilities as well as new possibilities.

I found all of this fascinating and brings my thoughts back to the Simulation Theory that I talked about in Chapter 1.

Today, we can play computer games such as GTA for the PlayStation and the Sims for instance. In these games, you will find computer-generated characters within any city, who live their lives as normal.

The gamer is able to interact with the characters, but when they are left to their own devices , they take a nap, eat, work with their families, and generally, live their life by using artificial intelligence which allows entirely random actions.

There are also games that permit you, the player (by wearing headphones) to be immersed in an immersive virtual reality as well as interact with fellow players and also computers-generated characters that are based on artificial intelligence software.

Although they're virtual reality games, and you may be able to fool your eyes however, you're always aware that you're playing because even with the finest graphics, everything looks cartoonish. But, that being isn't the case 50 years ago when we couldn't even imagine the technology we enjoy today , and as it's developing, we can think about what it'll look like in the next 50 years.

So, let's jump one thousand years now and think about the way our virtual reality computer will appear like. Graphics will look so amazing that they'll be just similar to what we see today Consider this: you turn your game and transform into an enchanted character. You could play as any person from any location and create an empire and life and even run the Government and hack into the game to become an elite and be in control of all things.

The computer-generated characters that go through their daily lives are free to expand, explore their surroundings, and even engage in virtual-reality games. This brings my mind back to chapter 1 in which I spoke about the numbers: 8 billion people on earth (in games) as well as millions of them engaging in their own game. There are eight billion playing their games, and in each of them eight billion, and so on. It's a lot of artificially intelligent people living their lives in a world that appears like, tastes and even feels as the one we live in today with the desire to discover, grow and expand, yet while being subject to rules and being controlled by unimaginable forces. Sounds familiar?

The Government sets the guidelines that we all adhere to and the fundamental written laws are easy for us all to study and follow. The way in which we are governed is less apparent to us but becomes clearer as our education/intelligence grows, in a game this would be levelling up.

If you see your life as a game you can participate in it by deciding on your own goals, and the purpose of the game is reaching these objectives. In other words, if life's something that is a game, and it appears like, feels and plays exactly as a virtual reality game How can we tell that we aren't an artificial intelligence programed into a computer that is able to go around our daily lives? What would we know?

So, what do we do? be doing about it?

The simple answer is no at all, but as individuals, we do have options.

If we're playing the world of virtual reality or not it's the same. Elites will forever remain the elite and the state is a part of everything , to the point that it is impossible to stop it.

But, just being aware that there are safeguards that are in place to regulate the actions we take, we are more aware and take risks to take our own decisions by recognizing the areas where we are being affected, and begin to think independently and educate ourselves, then only will we be able to influence change, and only then can we be truly liberated.

Chapter 6: What's The Reason

You're Here?

When it is born, every soul is bound by an agreement of sanctity in the Universe to achieve certain goals. It makes this commitment with all the fervor that it is. Whatever task your soul has decided to, all your experiences will help you to awaken the memories of the contractand make you ready to fulfill it.

--Gary Zukav

The Soul's Seat

Did we enter the physical world with a specific purpose or goals to achieve according to Gary Zukav in his quote above? Many believe that. If this is true, a part of us would need to exist prior to our birth. Let's take a look.

According to one Scientific Materialist each individual human is a collection of components similar to my twenty-year old Toyota Land Cruiser. Humans are made up from a heart, brain blood vessels, and muscles. The Toyota features fuel injectors the crankshaft, pistons as well as valves. It was built in the factory. It was assembled while during the birth. The Land Cruiser I've taken good care of it and it's outlasted most of its rivals and had a pretty good life in the SUV category.

Human beings too appear to be affected by the same forces of fate. You could be fortunate and have an educated, wealthy familia located in the West or be unlucky and end up in into the world from Somalia, Afghanistan, or any other Third World nation where living conditions are bleak and opportunities for a better life are almost nonexistent.

Is it just luck? We can tell from our examination of the evidence that the human body isn't a collection of components. What lies behind the scenes, supporting, and giving life to the body is a thought-based construction, an collection of memories that are through the medium of mind or the spirit. In his book A New Science of Life A Theory of Morphic Resonance (Park Street Press 1995) the

British biochemist Rupert Sheldrake, presented a theory that is plausible, based on the theory that is presented in the book. He claimed that our physical bodies are reflections of our morphogenetic field which is a combination of that of our grandparents as well as others in our species. The morphogenetic field that results is the one that shapes our body when we are in the womb. Genes create and release the required proteins in the right time.

Our bodies aren't made from separate parts. Every part is one. Human body's capabilities were formed by the evolution and experience of our species in a unbroken chain that dates back to the beginning of existence on earth. We are all unique only in the sense that we consider ourselves to be individuals. On a deeper level, we are completely at peace with the universal mind that evolved us. As we develop spiritually and become more mature as we do, we'll begin to recognize our connection with all that is created. If we are willing to answer the call to explore when it is offered and experience growth, it will be a process that will eventually lead to the perception of the world. The shift could be followed by an unexpected realization, or epiphany.

When a person who's experienced this understanding becomes wiser, goes through life, and progresses into middle age the notion of fate being whimsical and random will become increasingly untrue to what they have experienced. In the Nineteenth Century German philosopher Arthur Schopenhauer For instance, he stated within one of his works that once an individual has reached an age of maturity and reflects on the course of their life and sees that it appears to have been a continuous sequence of events as if written by the master of storytelling or novel writing. Particular events and interactions of people that appeared like they could been a result of coincidence are later discovered to be crucial elements in a continuous story.

If that's the case as my own experiences suggest that it is, then we have to inquire who was the author of the story?

Presently, Schopenhauer would have said that it was a person's subconscious mind. He'd point out that dreams are created by a part of us which we're not aware. He'd suggest that our entire existence is shaped by the subconscious part of us which he referred to as"the "will in us." The will within melds with the lives of other people so that all of our human experience is as the sounds of a symphony.

We now know that according to the theory, there is only one single organism with a mind is present and that mind is living the dream that which we call real. It is difficult to comprehend this since we are all an element of the whole organism. You I, myself, and everyone else are all participants in the dreams and each of us has distinct perspectives. Our perspectives differ depending on the position we're in with respect to the overall. This is the reality that which our minds subconsciously create. However, our own souls or subconscious minds are a part of the larger picture that includes the Infinite Mind, also known as the Big Dreamer.

This concept isn't new. Over 400 years ago, John Donne wrotethat "No human being is an isle." The lives of all people are interconnected. As part of humanity's continent We have roles to take on that affect other parts of the continental. A person we meet randomly becomes a significant character in the tale of our lives, similarly, we are key players for other people regardless of whether we are aware of it.

What is it about us, our own puppeteer with its distinct perspective that makes us want to perform our various roles in the vast human dream? What is the reason we aren't conscious of it? Did this component develop in the time the egg, the sperm , and the fields of morphogenetics that belong to our father and mother were joined by our own personal physical field? Did it grow in the same way as our egos evolved as a kind of parallel creation? A subconscious part of your conscious brain is able to store the memories and the programming that has been accumulated throughout your life. The other aspect of ourselves that is not in our conscious consciousness can be described as our unconscious mind, or soul. It is our personal changing field, which has developed through our personal development that started when our lives began. This is the part that we play as. It's part of the grand dream. It's a participant on the field of existence. Its purpose is to grow and development because it is one with the Big Dreamer.

Perhaps youtoo may have had to take a choice that was based on an ego need or arousal, the person who is scared, that craves to rationalize and wonders what others may think. Somewhere in your mind, you knew you'd regret your decision if you did not follow through. In that moment, you were in contact at a very brief moment with your subconscious mind.

One mind is present and is divided into several grades such as the Big Dreamer's mind is the largest, covering the entirety of what is known as mind, our collective unconscious, which we share with the rest of humanity. Individually, our subconscious minds or souls. Then, we have the conscious mind, commonly referred to as the ego brain, with an unconscious portion that contains memories and the programs for this particular lifetime.

Our conscious part in our is our objective mind which is self-aware. This is what trick us into believing that we are different. However, we aren't. In our own viewpoint, life might appear chaotic and chaotic, but everything is orchestrated on the level of our subconscious. We all have our own roles. The things go according to plan when we go with the flow. But when we wander off course by refusing to accept the phone call, things get awry and things get messy.

Maybe you know someone like me who got married to someone who was certain that they were going to divorce. For me, that sensation didn't come to conscious consciousness until after the invitations were sent out. But there was plenty of time for a call to end the wedding. However, he chose not to. After two years, his wife and he were separated He came to the realization that his ego had pushed himself into committing to the wedding because he didn't be brave enough to inform the girl, his friends as well as his parents and her parents that the wedding would be an error. It's a typical anxiety of the ego. If he had listened to the inner voice of his own then he wouldn't have been a victim to the terror that followed. The ego mind blocked communication by using the mechanism of denial since the truth wasn't what it desired to hear.

Elisabeth Kubler-Ross, born in 1926 andThe Swiss-born doctor and author of the best-selling bestseller On Death and Dying, was at the event and was able to in the healing process of hundreds of patients. She researched the near-death experiences of many others. She talked about her own mysterious experiences of being out of body and is generally regarded to be one of most respected authorities on this subject. She concluded that the inner voice is real. In a lecture she gave in 1977 at San Diego and published the following summer in the Co-Evolution Quarterly, she said, "If you listen to your inner voice, to your inner wisdom is more powerful than any other person's as in your case--you won't be in a hurry and you'll know the best way to live you life." It's unfortunate that my friend had not heard these thoughts, or, in the event that he was, they were not taken seriously.

How do we contact you? In answering the desire for us to become "born by the Spirit of God." Recognizing that we are component of the entire, and that we have access into the inner mind of entire inside us. When this is realized in us, it is time to are able to move to the Kingdom of God - the state of being aware and recognizing our connection on a gut level. As time passes, our ego as well as our lower-self will join in perfect harmony to our unconscious mind. It's inevitable and, when it does, we will be able to experience the biggest benefit of our efforts towards the spiritual path to a life where all the pieces are in place, and where we can understand what we're doing on this planet, what our mission is and how we can be successful in achieving it. The fear will diminish and eventually disappear. With time we become calm at peace, calm, and peaceful. We be in the present moment and, maybe in the very first instance, fully to be able to enjoy and live within the world of physical reality.

We all have egos. this part of ourselves that evolved in the course of our lives from a lack of focus during our early years at the age of the crib, to the part of us that holds the memories from this life. This it is that part of us which is worried about life, the part that fights and achievement, seeking fame and recognition. However, the subconscious mind or soul isn't focused on the physical world's trappings. It appears to have been there for a very long, longer time, dating back to the time of humanity's transformation from a race that was driven by instinct into an animal that is that is self-aware as well as free-will. It is not afflicted by stress or anxiety since it knows that it will exist for the rest of time. It has no need for self-importance. This is in line with The theory that morphogenetic fields are morphogenetic fields developed by the Rupert Sheldrake. Life itself is the morphogenetic field, which first separated from the rest of the field as DNA evolved into single-celled organisms. This field changed and evolved throughout the years as life took on increasingly complex forms. Through the years, various sections of fields took different pathways of evolution. The field I am in and then changed into your field follows the evolutionary path of primates. Every one of us has some part of it, literally, because it is all as television transmissions. The first time we experienced this is when we began to differentiate ourselves from other primates through self-awareness. Your field, sometimes referred to as your soul continues to grow in the same way that it is reincarnating time and time again.

Apart from fitting with this theory the evidence for reincarnation that has been compiled by researchers like Ian Stevenson and Jim B. Tucker from The University of Virginia is simply enough to make it impossible to ignore. For those who are religious Christians might be put off by this or simply dismiss it as a matter of fact because it's not a part of the doctrines in the Catholic Church. But I am convinced the fact that Jesus and his contemporaries - both Jews and pagans took reincarnation for granted, much like Hindus or Buddhists do nowadays. Go through the Gospels with this thought in mind. You'll see certain passages which were previously thought to be unimportant are now clear to your eyes. Many scholars believe that when church canons were codified at the time of the Fifth and Fourth Centuries the idea of reincarnation was deemed to be unhelpful. The idea was that prospective converts might be resistant or put off acceptance of Christ due to the belief that they'd have more opportunities in their lives in the future. Reincarnation was eliminated from the Bible in the wake of.

Consider a few instances which suggest that Jesus and other people at the time believed in the idea of reincarnation. For instance, John the Baptist was widely believed to have been the prophet Elijah Reincarnated. Jesus himself claimed that this was the case. (See Matthew 11:14.) A few times, Jesus asked his followers whom they thought that he (Jesus) is. They answered that they believed in him (Jesus) as a prophet, possibly who reincarnated as the last prophet was around 400 years before him. Take a look at the story of Jesus healing the blind man in the passage in John 9:1-12. The story is with the following words:

As he walked as he walked, he saw a person blind since birth. His followers asked him "Rabbi, who did you sin the man, or his parents that the blind man was born and blind?

"Neither this man or his parents committed sins or committed sins," said Jesus but this occurred in order that the will of God could be manifested by his character.'

Because this man's eyes were blind since birth and was born blind, the only way that his sins might have contributed to the blindness of his was to have committed a sin in a previous life. Jesus did not inform his followers that this was not likely to happen. On the contrary it appears that he believed it was possible however, he also gives a second reason why the man was blind.

When I've been researching reincarnation, discovered that libraries are with books with information on the topic. After becoming interested in the topic, I've encountered and become acquainted with several people who earn a living helping others to remember their previous lives, and then let go of the buried memories that are keeping them from reliving their lives. In certain cases the passage of time has been thousands of years been passed since the debilitating event was triggered. I've been to the School of Metaphysics in Missouri and watched the trained people who read the Akashic archives report on the previous lives of workshop participants. In addition I've read four works written by various past-life psychotherapists, and edited the fifth. Instead of describing what's found in these books documents, I'll provide a brief summary of an incident that was reported in the book of 1988, Many Lives, Many Masters. I've picked this book since the writer, Brian L. Weiss, M.D., cannot be accused by anyone as being an actual Looney Tune. He is an Phi Beta Kappa, magna cum laude graduate from Columbia University who received his medical degree from Yale where he interned at the New York's Bellevue Medical Center, and was later appointed chief resident in the department of psychotherapy in Yale University. Yale University School of Medicine. In the time of the incident he wrote about in his book, he served as director of the department of psychiatry in Mount Sinai Medical Center in Miami Beach.

Weiss is a doctor and scientist, who has published widely in journals of professional excellence. Being ethnically Jewish He was skeptical and did not believe in Reincarnation. He was aware that the vast majority of his colleagues in the field don't believe in these things. He was patient for six years before giving his faith to the notion that he was obligated to share what he'd discovered. He was more likely to lose than gain from telling the story of a woman named Catherine (not the real title) who was referred to him in 1980 in search of help with anxiety, panic attacks and fears. Check out the book. I'll just touch on few of the points.

For the next 18 years, Weiss used conventional therapy and Catherine. This means that He and Catherine were able to discuss and analyze her life and relationships. When nothing was working He tried hypnosis as attempt to discover what she might be suppressing which could be the cause of her anxiety. Unremembered events from her childhood In actual fact, they were exposed that appeared to be the source of a number of her issues. As is typical in this type of therapy she was told to recall these events after being removed from the state of hypnosis. The doctor. Weiss discussed what had been discovered in an attempt to alleviate her fears. As time passed her symptoms remained the same as they had before.

He tried again hypnotism. The second time, he brought her all the way to her age of two. However, she didn't recall anything that shed any new light on her troubles. He gave her clear directions, "Go back to the moment when your symptoms manifest." Nothing could have given him the information he needed for what was to happen the next day. She was able to slip into a previous life that occurred around 4000 years ago. Weiss was amazed when she detailed herself as well as her surroundings and other people during that particular time, with particular episodes and later whole lives, which appeared to be the root cause of the problems. All in all, she claimed she lived at least 86 times. This, however is a sign that she was a young person considering that the majority of people patients who have been through this kind of therapy have gone through thousands of times.

Weiss continued to use hypnosis in attempt to eliminate Catherine of her nervous disorders. Through weekly sessions that lasted several months, she was able to recall and reminisced in detail about the best moments of her previous lives as well as the moment that she died in every. People who were in a position in one life often reappear as a different person in another one, such as the Dr. Weiss himself, who was her teacher 3500 years prior to her death.

Catherine was not an enjoyable life over the past 40 years. The vast majority of her memories of her past experiences were painful and were the cause of her current ailments. Making them conscious and discussing them with her helped her to heal. Given the severity and number of her neuroses, therapy typically would have taken several years before she was healed. However her symptoms vanished within a matter of months. She felt happier and more peaceful than she'd ever been.

Weiss is a highly experienced psychotherapist who has treated hundreds of patients. He is convinced Catherine wasn't a fraud. She was uninformed and averagely intelligent an untrained young woman who earned her living as an laboratory technician. He believes that it is quite unlikely that she could pull off such a slick hoax and continued to do it every month for months. Consider it. She was attractive physically 28-year-old woman of average intelligence. She was a high school diploma as well as some vocational schooling. Could she have been able to fake her neuroses? Could she have faked a gradual improvement between visits until the following, and eventually to the point of being totally unaffected? It's not likely. Furthermore in this instance, where the plot gets more complex She shared details about her father and a newborn son and a baby son, both of whom had passed away. Weiss believes she would not have learned anything about them from the normal channels.

The message that comes from another side points to what may be the most fascinating aspect of her tale that is the space between her the lives of her past.

After being killed, she escaped of her body and reincarnated extremely quickly. After her second life, she recalled the experience as like the one described by thousands of people who were diagnosed dead only to came back to life. She emerged from her body, was at peace and conscious of an energy-giving light. It was during this time that the spirit guides spoke in a conversation with the Dr. Weiss for the first time. In the loudest, most husky voice, without hesitation, Catherine declared, "Our task is to develop, to be God-like by learning. We are apathetic. You're here to serve as my teacher. There is so much I want to discover. With knowledge, we are able to are able to approach God and after that we can take a break. We return to help others."

While Catherine could recall her past lives once she was released from an hypnotic state but she was not able to remember, nor was she interested in recollecting what conversations. Weiss had through her with various spirit entities. The "masters," as he called them, communicated with her in a way that was primarily for his own benefit, and not directly for her benefit. I won't discuss in detail these conversations; you may be interested in reading the book. In essence, they said that we are incarnated into the physical realm to acquire the things that are not taught on the other plane. In this realm, anything that you feel or imagine is instantly real or magnified. Any slight ill-will towards one person can turn into anger. A simple feeling of affection transforms into a complete love. If you think of the demon, a mental that resembles it will appear before you. If you imagine in your head a gorgeous sunset that is viewed from a quiet beach, you'll find that you're there. It is due to this that we require the thickness of matter. Matter slows down things to allow us to figure them out. Earth is a classroom. The most important lessons that we learn here are hope, charity faith, love and faith and to be able to trust and not be afraid.

Let's go away from the Dr. Weiss for the moment and explore the mechanisms of the reincarnation process. What you're about find out initially seemed incredible to me as it could be for you. However, just like I do, like Dr. Weiss, I feel obliged to give it away .

If you experience a Life Force or spirit is removed from any object whether it is an animal, plant, or thing The Life Force continues to exist but the object it was supporting is no longer controlled by it. The thing ceases to exist, influenced by the Life Force--and then goes into dust. This is what happens to the things we typically consider to be living things, such as animals and plants, as well as this is also the case for things that we thought up to that time, were inanimate objects like rocks, moons mountains. While the decay process and the returning to dust is slower for the former however, it is still happening at the point that there is no Life Force is no longer in existence.

According to what Claire DuMond came to know in my novel The Secret of Life: A Journey Out of Body into Mind The secret to existence is "urge to evolve," which in this book I've referred to as"the Life Force or the subjective foundation of the mind. It encourages development and is the opposite of the process of entropy. The way I see it the life force is the ability to organize, which pushes it to develop into ever higher shapes. Then, when the sense of being separate is experienced the soul, or a unique subconscious mind is born.

Your subconscious mind , or soul could be evolving in the past on Earth or be evolving elsewhere. For me I've experienced flashes of memory that could be described as an epiphany. It is something I believe to be an indicator of my personal story of development. This flash of memory was perhaps 30 seconds long and the "relived" the entirety of my prehuman existence in rapid succession, starting from my life as a fish-like creature living in the sea of the past, to reptiles, to an animal-like furry creature which lived in the form of a tree. This is a good fit. Once a soul has experienced all it can about itself in one form, it is seeking an experience which will enable it to push forward. In the end, it will expand and grow until it attains an ideal state.

We are living in a multi-dimensional world even though in normal circumstances our physical senses enable us to feel only width, height and the passing of time. Souls are evolving in different dimensions, and developing on other planets in other solar systems of the universe. Even though you recognize the possibility that your soul is more advanced than the life you live on earth, one perspective of how souls developed along with our planet will be discussed in the next paragraphs. This isn't intended to be the definitive view on the subject , and is only intended to give you a plausible explanation of the evolution path souls may have traveled.

One of the theories that has been accepted by a few adherents of the

Eastern faiths is that souls that embarked on their journeys to this world were present in some form or another since the time they first came to Earth. They did not separate but they did so until the time of the legend of Adam and Eve. Scientists are likely to estimate that this happened around 200 million years ago. As we mentioned it was the time the time in the development of our minds at which we began to be aware of ourselves. We saw ourselves as distinct and separate from all the other creatures. In contrast to animals and birds in the forest, or the Savanna, we never depended on nature or our instincts to dictate our behavior. Our minds could alter what our instincts told us to do.

This is what the tale about Adam and Eve is about: the growth of objective awareness and breaking off or dissociation from the field that brought about self-awareness as well as the freedom to choose. God advised Adam as well as Eve that they should not consume from the tree of wisdom that reveals goodness and vice versa. The snake, symbolizing Adam and Eve's all-too human nature also known as the ego brain, told them to take a bite. Instead of consulting God prior to making a decision, Adam and Eve acted like we behave today and then proceeded to do whatever they liked. In exercising their the free will of this way they cut off their connection with God and humanity is suffering ever since. We've, in effect eliminated ourselves from the Garden which means that we're no longer in a position to easily tap to the abundant nature will always provide us with.

The best way to go back is to reconnect to our subconscious mind, also known as the soul, and consequently our connection with the higher. This isn't an unrelated topic. The main point is that the genesis and development of souls in this particular situation could have been a continuation of the process of evolution, from animals with one cell in the ocean, to creatures who first moved on the land, then to tiny mammals, pre-apes, and finally to homo sapiens. As homo sapiens that we began to be distinct. This Adam step and Eve action was essential. However, we've been in a plateau for over a hundred thousand years. Because the evolution process is more of a spiral than straight line, the next step is to go back to the condition that which the first couple had, only on the higher levels. It is time for the majority of people on earth to connect with the universe and yet remain aware of their individuality and retaining the freedom of choice. Metaphorically, they will go back in their place in the Garden with a fresh perspective and, remember that they must keep God on the inside. In the event of this the worst times will pass. Every desire we have can be realized.

Some people will claim that reincarnation isn't possible since there are more people living now than there were before. The earth's population has increased exponentially in recent decades. If humans aren't human without an soul, and if humans' souls must be developed through multiple incarnations, then from the question is where did souls get their start? It is my opinion to me that there could be at most two possibilities for theories. One possibility is that souls that developed on earth are incarnating more often than they did before. That is they are spending less time between lives. One therapist who assisted many clients in overcoming mental health issues resulting from past life events has indicated that, with regard to his clients, the length of time between lives varies between 800 years to a minimum of 10 months.

Another possibility is souls are pouring into the world from everywhere The development of souls to now may have happened elsewhere for a lot of people living today. Morphogenetic fields consist of information , not energy. Based on quantum physics unlike energy which must travel and decreases in intensity as distance increases the information is everywhere within the field simultaneously. It is not local. (This is also, of course is in agreement with Thomas Troward's idea of thoughts and the spirit or life which was discussed in a prior chapter.) Since a thought form is everywhere at once--nonlocal--souls can come from anywhere in the universe; no travel time is required.

In a way, we can anticipate the life forms that have developed on different planets will have similarities to earth's as long as the physical conditions on the planet are comparable. This is the case for life forms that developed in Australia when the continent was separated from the globe's land mass. We can see the equivalents of dogs (dingoes) and cats as well as other animals beneath that aren't precisely cats or dogs. They are marsupials and not mammals. If you were to look at the family tree, they'd be closer to the opossum, rather than the animals that they resemble. Water has impeded the spreading of genes between dogs and cats, however, not the morphogenetic field.

They are everywhere, unhindered by water. These fields have affected the appearance and shapes of Australian animals, though their genes could be very different. If you think about it, those bizarre creatures could also have evolved on a different planet when physical contact with the globe was cut over many millions of years.

Let's now look at the process of development of the human soul. When a human soul that is new begins its journey, the amount of naive or wrong actions, thoughts , and words the soul is accountable for is much greater than that of the right kind. This is to be expected. This is also the place where the law of Karma enters play as a primary teaching tool offered to us by our universe. Based on the law of karma which is a simplified version of the principle of cause and effect any thought, thought, or action must have a certain result, either good or not and the outcome should be experienced by the individual accountable. Learning about the law of karma is just one of the ways to learn. It's one of the reasons that numerous lives, both physical and spiritual, are required. It is the case that we don't remain long enough to allow all of our actions and deeds to unfold in a single lifetime.

The Bible states, "As you sow, the same will you take," or as an old friend of mine from advertising was frequently said, "What goes around comes around." I think Jesus was speaking of the law of karma for instance when Jesus stated:

Don't judge and you won't be considered to be judged. Do not judge and you won't be judged. Give forgiveness, and you'll be granted forgiveness. Give you will receive it. yours. A decent amount, when pressed down, shaken and then poured over it, will fall in your hand. With the measure you select it will be measured to the size of you.

-Luke 6:37-38

A self-centered act by you that causes pain to another person earns you a piece of bad Karma. This should be compensated by suffering as a result of another similar act by someone else, whether during this lifetime or in the next. Your kindness by you earns you the benefit of good luck. The outcome of this act could be the elimination of a bad karma or exactly the same level of generosity from someone else. One could consider karma to be the metaphysical law similar to Newton's law of physics. For every action, there's an equivalent and opposing reaction.

The moment I learned of this and realized that it became apparent I began to think about my life in the past and recall my actions towards other people that caused them hurt. A few instances of naiveté, as well as three or four instances of outright brutality, came to the forefront of my mind. I was truly remorseful, and fell into a state of depression. It was like an unsettling cloud of darkness hung on me, and I was unsure of my life. I could feel the hurt I'd caused, and was contemplating how I would ever pay the loans. At the time I didn't realize the therapeutic value and practical benefits of making a direct confession to God or Christ and asking forgiveness. In fact, I believed that I was destined to endure the same level of suffering as I'd caused.

Let me add that this took place over thirty years in the past. I've come to understand that karma isn't an act of punishment or justice. It's a teaching tool. When a person is taught the essential lesson and has a positive karma, that karma will be removed. This brings us to the topic of an individual God and the force of creation I've called the non-dual and subjective foundation of mind. Edgar Cayce's writings suggests that there are both. According to me, our personal God could actually be our soul's evolving self or Higher Self.

Let's return to my question of how to pay back my karmic debts, which I briefly mentioned a while ago. A few days ago I was walking along the Canal of Burgundy during a trip to France. I stopped and yelled "Please, God, Please. Make sure that the score is even. I'd like to play on a level field. Let whatever you can occur to ensure the obligations will be paid."

A Jewish friend of mine later informed to me Jewish prayers to clean the slate include the following phrase "But not through suffering or pain." It occurred to me, "Now he tells me." I've learned for myself that you will receive exactly what you request God for when the request you request will lead to spiritual improvement. What I wanted started to manifest three days later , when I returned to the States. I brought my daughter's brand-new 10 speed bike out to a spin test across the hill that is right in front of the house. My foot got caught on the road and the pedal completely cut the Achilles tendon in my. It was a horrifying injury. I was in the hospital. I had two surgeries, experienced an awful amount of physical pain and was placed in an incision from the tip of my toe all the way to high thigh area for eight weeks. It took me nine months before the wound was closed completely and then another nine months before I could walk with no limp.

But that's not the end of the story. As I was still being cast in an apron, my wife declared that sparks was gone from our relationship. She was fed up from living in America. United States. She was moving out of my home and filing for divorce and she was taking my daughter with her to France.

As as a parent, I imagine that the loss of a child could be the most painful experience that one could ever endure. If this is the case the possibility of having your sole child of twelve years old relocated to a distance of three hundred miles and six times zones apart is number two. It wasn't a very good year but at most, the slate was washed clean. Like you would expect I grew spiritually. Adversity is an excellent teacher. It also made me realize that it's possible to communicate to that Big Dreamer and get the request accepted. I don't advise you to take the same route since I believe that the process of achieving an equal playing field regards to karma doesn't require a lot of pain. This is due in part because the goal of karma, which is the principle of cause and effects, is not retribution in itself. The Big Dreamer does not find satisfaction in obtaining "an eye for an eye and the tooth for the tooth." Instead like most things that are connected to consciousness and law of the universe, the aim of karma is the promotion of spiritual development. Sometimes, an "eye to one eye" is the only way to prove an impression. This is particularly true in the beginning in our journey to spirituality. In the end, the only way to be able to fully comprehend the implications of our choices, thinking, and our plans is to experience the consequences in person. If O. J. Simpson returns in a different life as a woman only to be brutally killed by a larger and stronger man with knives, it is likely that you will eventually "get the message" inside his heart.

The ultimate purpose in the law of Karma you'll notice, is shifting consciousness I've been talking about. It's an "aha!" experience, the moment of realization at the core of what has been done when one realizes that they as well as others are one in the same, not literally right now, maybe however, at some other momentor other version of. It is probably the thing Jesus was aiming at when he embraced the child into his arms and told him, "Whoever welcomes one of these children in my name will receive me. Whoever welcomes me is not welcoming me, but only the person who invited me." (Mark 9:37) When he spoke about the thirsty, hungry and the disadvantaged He said "I say to you, I am that everything you've done to the least brothers of mine you did it for myself." (Matthew 25:40)

In these passages, Jesus is saying we are all of God and are all part of God We have all played, or will play, every role and contribute to the creation dance. That being said helping or harming one another is to assist or hurt yourself in the past or in the future. It also helps or hurt God.

How can we assist or hurt God? Only one screen of consciousness that the moving image of the real world plays. In total, the entire screen is God's consciousness. The awareness of each person is, however is only a tiny portion of the bigger screen.

Do those who have committed murder in the present or previous life be subject to the same retribution? I believe that the practice of meditation and reflection, as well as study, can help to attain a higher level of consciousness. This, when combined with repentance and forgiveness, could avoid having to suffer "eye for eye" punishment. The subjective mind of the universe isn't judging and does not hold grudges. Being subjective, it can't ever sit down and consider. If retribution isn't required to grow and development, it won't occur. That is one of the main message of Jesus. Jesus came to reveal the way to Christ-consciousness, and to eternal life. The moment of awakening to

Christ-consciousness is the trigger for an event that is known as the "law of grace" which eliminates the necessity for an cosmic boomerang. Spiritual awareness "fulfills the law" to borrow Jesus"law of grace," in that it shatters the false beliefs that were the root of the wrong decision. "I am not here to destroy the laws of God," is what Jesus might have stated "but to show you how to follow it through a spiritual awareness."

Achieving this level of consciousness is not an easy task. "Remember," Edgar Cayce declared at the time of one of his books, "there is no shortcut to an awareness of God's force. It is part of your consciousness however it is not attainable simply by having a desire to be there. There is often an urge to desire to be able to attain it without applying spiritual truths by means of mental processes. That is how you go to the gates. No shortcuts are available to metaphysics, regardless of what the claims of those who experience visions, interpret numbers or look up the stars. They might encounter urges however, they do not control the will. Life is learned by the self. It's not something you proclaim but you do study it."

What exactly is it that one get how to do it? Regular prayer and meditation. Fasting. The study of the Scriptures and books similar to this. Giving service to peers. These are all beneficial. Depending on where an individual is in their evolution regular application of these practices could be enough. Personally, I've been doing it for forty years in a focused manner and can state with absolute conviction that I'm still in an excellent path to take. Sometimes, when I sense that things are in a negative direction, I go through moments of doubt and fear. From an intellectual perspective, I know this is not beneficial. If I was fully developed I would not have these doubts. However I'm using my talents and am content in this way. I have a strong and comfortable marriage that is loving, warm, and cozy. I take pleasure and share in the happiness of happy, healthy and well-adjusted children. My biggest weakness appears to be the worry that if I do not be a bit more active, the positive things that the universe is pouring towards me will eventually cease to flow. Perhaps it's due to being poor in my youth. But, I'm getting better. I'm not perfect and I'm aware of them. Many would say you're patient. I'm sure I'm too easily angered. I'm working to change this. If I continue to work on it, perhaps achieving full control of life might not be very far off. What is this life? The following? The one following that?

I am convinced that I been born with a degree of development greater than I currently am able to achieve. However, for the first part of my life during this time I was sliding backwards and losing ground. After coming to the realization that I'm not alone I've been able recover a lot of the ground lost quickly since the only thing I've needed to do was "remember" (re--member) the things I knew already and regain the consciousness level.

Backsliding can be a risk the old soul has to face as it incarnates. However, as is the situation, the consequences of the unwise exercise of freedom can be utilized for the Big Dreamer to provide an opportunity to grow spiritually. The challenge of having to go back to the beginning, for instance, made me more prepared in writing this novel. If I'd come into flesh without prior knowledge I would have had two issues when communicating this information. The first is that it would have been pre-existing, or a component of me or my identity, making it difficult to express. The second reason is that it wouldn't have been acquired in the blaze of a world of skeptical people and doubters. However, since it was gained slowly over time I am able to provide a solid understanding of what arguments must be overthrown. In the end I had to confront them by myself before I accepted the information as factual. This may make my argument more convincing.

After regaining what I am able to offer in this life and I am able to do it within the context of this lifetime and the moment in the history we are part of. My heartfelt desire and intention is to help you as well, to remember your past experiences and to assist you in your quest to climb to a new level of knowing. However, let me note that growth can't be caused by force. If the heart is tender enough by charity, the result like this, for instance, will be Apostle Paul's phrase "as as tinkling brass." The wealthy man who gives his money away in the hope of purchasing his way to heaven hasn't bought anything. It's the man's heart that matters. What he really gives is kept due to the happiness he experiences. But don't misunderstand. It doesn't mean that he must not do something, even if there is no joy in his heart. Maybe his act is what his soul is looking for to get it started in the right direction. However, this will not allow him skip a class. Spiritually speaking, can't leap into college.

Let's consider for a second about the soul that is emerging with no knowledge. In the beginning the entity will go around causing a lot of trouble. For the most part nobody is likely to endure more hardship in a single lifetime than they are able to endure. The bad karma of the past that can't be cured by good actions or poking the lips will be stored to be dealt with in subsequent lives.

Think of someone you know or came into contact with, whom you consider to be a young , innocent soul. The person seems to be apathetic and naive. The person simply doesn't hear the voice that is still a whisper and doesn't think twice about spraying their house with bullets for the pleasure of watching broken glass shatter into the ground. In the beginning such a person will accumulate more in karmic debt than they can pay off. As time passes and the person continues be incarnated and incarnate, the connections between his subconscious mind and ego expand. When the communication lines are opened, the conditions improve. Finally, the correct frequency is identified so that the mom ship is heard through with a loud and clear. This is the reason why someone who has a long-standing soul has an advanced sense of what is right and wrong. If this is the case, bad karma that has been accumulated from earlier lives is taken down. Many of us are in this situation today. Perhaps you're one of them. If yes, a quick jump is all it takes for that change in your consciousness.

This book is not for people who have young souls. Therapists from past lives tell me that a lot of people come into this world with no plans for their life. The lives they lead are chaotic, and in the worst case, void of purpose. They're unlikely to be able to tap into the vastness of the universe and it doesn't seem likely that they'd be able to show much curiosity about the book's say. But I'm sure that if you've stuck with me to this point, I believe that you're a fairly old soul and are close to achieving making a breakthrough. If this is the situation, it's very likely that you came into this world with unique talents that you can use to benefit humanity and you have a significant objective or purpose to achieve. If you've been able to attain Christ-consciousness in the past and are now on earth, the purpose of your life is to assist others to achieve the similar. If you're still a long way to go, it's to confront and conquer some issues or resolve karma from an earlier incarnation. We'll soon be discussing how to get rid of the karmic baggage that is preventing you from moving forward in order to get into the flow.

In terms of flow I've heard a lot of people who claim that they are New Consciousness thinkers repeatedly make assertions that "everything is functioning exactly as it ought to" at any moment. Perhaps they are trying to remind them not to focus on the outcome instead of focusing their efforts on the task they're here on Earth to accomplish. I'm in agreement with this method. I've never earned money, for instance in my efforts to earn money. I'm sure there have been instances that I've tried my best. In reality I've made lots of money by focusing my efforts on the things I excel at as well as "just taking it to the next level," to borrow the Nike slogan. When I've let the chips fall in any direction it was, the results have turned out for my advantage as long as I was doing the things that I'm here for.

But I don't believe that everything works out to its best in any time. Although I don't often speak up about it, such assertions often irritate me since they violate the law of freewill. On the physical plane, also known as here on Earth every human being has the freedom to make a choice, and consequently, is in a position of making mistakes. Dumb mistakes. Incorrect calculations. Making mistakes in judgement. It is our right to be cruel or foolish. Thomas Troward pointed out that the universe operates according to laws.

Infractions to a law, regardless of whether or not you believe that the law exists and you'll suffer the consequences. If you're not convinced observe what happens when a shrewd child inserts his finger into an electrical socket. Metaphysical laws are as reliable as the laws of Physics. For instance, it is an established law that what you think will happen will occur eventually. Thus, believe that you're going to be bankrupt, and you will.

Wait a minute, you may say. I've considered that horrible things could occur a few occasions, but they never did.

The important point lies in the fact that "thought that it could occur." It's likely that you thought they would not. While it's not perfect hope is a form of faith. It's a belief in possibilities. You believed that this might occur or it could happen. You had sent your subconscious mind mixed messages and, fortunately for you optimism was more powerful and prevailed. You'd be better off to be confident that you'd succeed , and not to take risks.

Many people believe there's an angelic fairy godmother who watches over their needs, or perhaps an angel of protection It's fine. This could be the case. Actually, I believe there are guides or souls in the ether who are accompanying us on our journey of life. In a sense the spirit beings of these entities are in us, and are always present with us. But I believe that things occur because the law faith is at work. It's all about keeping constantity so that we're in a position to understand how the world works. But, I believe that "divine intervention" is also a possibility but it's not often, since it is a fact that, in the past I've been through what to me could have been anything but. It is only when the consequences of not occurring would negate Spirit's desire to grow and development.

I believe that faith could bring about something that appears to be divine intervention. For instance, Jesus is said to have transformed waters into wine been walking on water, and feed 5,000 people using just a few fish and a few bread loaves. Jesus did not just believe in his own words, he realized the presence of and the extension of his "Father on the other side of the heavens." There are personal accounts from friends I trust -- one of them was the founder partner of a major law firm--of miraculous events similar to those carried out currently. Making wine from water is just one of them.

It is important to recognize that the things you plan aren't guaranteed to be in order. Edgar Cayce indicated that things may go wrong, even if they've been designed with the clarity that is evident in the spiritual realm. Cayce assured us, for instance that birth and rebirth doesn't always work the way it is intended to and that there are times when mistakes occur. Cayce explained that we pick our parents and our situation according to what is available at any moment. The situation may be quite imperfect, but we can choose to continue regardless and, literally speaking we should hold our fingers in the air. The result could be that a soul will be able to discover after having made a decision and had been born, that its parents aren't fulfilling the promises they had set prior to birth. Contemplating that its personal motive for incarnation could be shattered by the new situation, the soul could choose to withdraw. Cayce stated that this could be the reason behind at the very least, some infant deaths.

Following the death of a person The soul's guides or teachers, or what may be called elders ranging from one to three according to the Dr. Whitton in Life Between Life-- will expose the recently returning entity to a comprehensive analysis of their recently completed life. They do not make judgements about the entity, but. The entity judge himself or herself, and is actually feeling the hurt or distress as well as the joy that they may cause others. Elders or guides offer comments and offer suggestions. They're non-judgmental in their approach and frequently give comfort to the individual that may be astonished by the report. It is likely to occur in the event that a few mistakes were made, or chances were missed that could have resulted in the fulfillment of the goals defined for the entire life.

This is a great place to share a few stories related to this. When he heard that I was writing this book, a close friend informed me about a situation that took place that changed his life while the time he was in his teens. He was unhappy in his home and was extremely depressed. His mother was crazy and she and his father were constantly at war. He had almost no friends. It was so difficult and the situation seemed so desperate, that he was considering suicide. After he'd been asleep with thoughts of suicide in his head his body, he felt the feeling that he was shaken. He woke up and observed two strangely dressed men. They grabbed him and dragged him up, straight through the ceiling as well as the roof. Or so it seemed. He is now aware that he traversed the tunnel we've heard about, and ended up in a sort of suspended state where three of them engaged in a discussion.

"Don't take it on," One of the men said.

"Do What?" he asked.

"Don't take your own life."

The man looked at him in a smug way and then frowned.

"It will do you no benefit," the other said. "If you fail to do so then, you'll be returned again and again until you've got the right answer."

My friend knew exactly what they were referring to. If he died prematurely, he'd be born under the same situation every day, much like the character from the film Groundhog Day, who had to experience the same day repeatedly until he realized what was happening.

Another acquaintance, a therapist who has an Ph.D. is using the concept of past life recall to assist his patients in getting over their fears and neuroses. He's gone through his own livesand claims that he's experienced the ability to go the past 15,000 years back prior to when his arrival on the planet. In that period, he involved in an intergalactic battle, which was a bit similar to Star Wars, I guess. His crew was captured before they set out to destroy an alien planet that was inhabited by the enemies. The crew and the captain were sent back to Earth for rehabilitation and imprisonment as well as rehabilitation. He's been here since then, being having to go through the cycle every time. He'd like to break free of the cycle, but so to date, has not been granted. He believes that he's part of a tiny portion of people in the world who are in an penal colony. For these people, Earth might be compared to Devil's Island in the South Atlantic located off on the shores of French Guiana that was an French penal colony from 1895 until 1938. As opposed to criminals of France However, prisoners from other planets are brought there to "get out of from the street," so to speak. He has estimated the number of prisoners in the world at around five million.

Recalling our explanation of the inter-life span, quite a portion of the time is in between incarnations, where an individual will examine his or her most recent life along with the other lives that he has had. As souls get more advanced, they are likely to have more time between incarnations than less advanced souls. They are more cautious when choosing the conditions of their upcoming lives. Additionally, parents who are compatible don't show up as frequently for more mature souls. While they wait the soul could make use of time to sharpen specific talents or abilities.

When back and on Earth souls can achieve the goals set for a specific incarnation more than expected. This was the situation for One among Dr. Whitton's research subjects. With nothing to complete during this lifetime the possibility of a premature death could be expected in order for the soul to come back for rest and recycle. Instead, the soul was assigned a new mission and was permitted to remain on Earth and to continue growing. In this way, the rest of the incarnation was successful and productive. What would be considered an untimely death for us mortals was avoided.

It's crucial to understand that taking on a brand new task is feasible. This means that we don't have to be dead just because we've achieved the goals that were set out before our birth. In essence, we could start a new life without experiencing the drama, stress and stress of rebirth, death the adolescent stage, childhood and so on. If scientists do find ways to unlock the mechanisms of aging, it could be the norm, provided that people realize that the goal of life is spiritual growth, both their own and those of others. This is because it is only logical to assume that people who make the decision to take a break from golf and retire every day, are likely to suffer from an illness, accident or some other incident to put them back to their normal routine and productive again. Let's revisit the lives between lives. If the time is right for rebirth, a jury of the judges go over or assist in identifying objectives and lessons to be learned for the next time around, and possibly offer the entity the option between a few different families that it can choose to join. The soul has to accept the decision, but it is apparent that this consent is typically given without hesitation. The new version is designed in the same way like a writer would outline the storyline of the novel. All the elements need to be set up and the supporting characters are ready and waiting. The whole process is difficult. The race, the nation, and the family's circumstances are all factors. Anyone who was a racist in the past may be able to return dressed in their race that they had been discriminated against to see the opposite side of the coin and work out the karma that they created.

The time gap between lives is vital. If we are aware of what's to come then the goal is defeat. Courage and fortitude as an example can't be developed in the event that the terrifying incident and the outcome are known beforehand. In order for lessons to be retained it is necessary for them to occur naturally and without prior knowledge. As the new ego starts to emerge the amnesia begins to set in. It starts as a child begins to bring the world around it to the forefront.

Many people who have been in the presence of babies and young children are aware that they've just come from a heavenly realm. I was able to see this clearly for my own son Hans who was in the toddler stage. If his needs were being met with food, as well as lots of good and healthy interactions with the people who loved his, he was happy to be there. His face appeared to be glowing. The beauty I see in my mind is appreciated by poets through the ages. It was most effectively captured according to me through William

Wordsworth in the fifth line in his work "Ode." The work is probably best known through the title, which is, "Intimations of Immortality from Recollections of Early Childhood."

The birth of us is just an unrequited sleep:
It is the Soul who rises up with Us, the life's star,
Hath had a different setting
Then, from afar:
Not in entire forgetfulness,
But not naked,
And it's through clouds of glory will we see
From God to God, our home:
Heaven has a plan for us to be when we are infants!
The prison's prison-house's shades start to fade
At the age of the boy,
He sees the sun, from whence it comes,
He is able to see it through his joy.
The Youth, who move daily further away from the east
It is necessary to travel, and is Nature's Priestess

And , through the magnificent vision
Is heading to be attended;
The Man sees that it is dying,
Then fade away to light daytime.

The extent to which a person "trailing cloud of glory" is able to make progress through this life, or if he achieves his destiny or slips backwards will be contingent heavily on the effort and capabilities of the parents. Think about the importance of ensuring to create a nurturing environment where children is able to flourish and grow, and discover "natural" talent. We could make it easier or challenging and, as a result, create an abundance of positive or negative Karma for our own lives. In a way our parents are our guides until we're old enough. It's a huge obligation.

All our invisible guides can offer is to provide us with direction when requested and assist in creating favorable conditions to "pursue our passion," as Joseph Campbell has said many times. Naturally, our unconscious mind will always remain a subset of the unconscious that does all it can keep communications lines open. However, each one of us has the freedom of choice. We have the freedom to go against what intuition or "better judgement" informs us.

One of the most important points of this chapter is you're here because of some reason. The reason you were born was not an accident. You have an option. You can look for this reason to live your life as you please, or choose to do what you want and possibly go lost that you'll never return. You're a soul in an actual body and not a body that has an soul. You may have made a few errors. If that's the case, it might never be late for you to fix the mistakes, particularly now that you realize that you're the driver, rather than the automobile.

It's difficult to deny that we're all evolving. Some do so rather quickly. Some take longer. Some might not make it.

What is the reason? What's the point of this?

In his book The Seat of the Soul Gary Zukav claims that the reason for this is the finalization and enlargement that the soul. Zukav wrote "When it returns home, the knowledge it has acquired during that time is evaluated by the assistance of loving guide and teachers. The new lessons which have been revealed for learning, as well as the new karmic obligations that have to be met, are examined. The experience of the incarnation recently finished are examined with a full understanding. The mysteries that surround it are no longer. Their motives, their causes and their contribution to the growth of the soul and to the development of the souls that the soul shared its journey are disclosed. What is balanced, and the lessons learned helps the soul to move ever closer to healing, completeness and integration."

Let's discuss the issue again. A lot of people believe that it's to be co-creators. But is that really the final point?

If we consider this for a long time, it may occur to us that the ultimate unconscious, the universal "one life" has begun reproducing itself. Take a moment to think about it. According to Richard Dawkins observed in his study of gazelles and cheetahs Propagation is an essential concept in the natural world. Each organism, from the smallest amoeba all the way to the largest whale is pursuing this principal goal. Why is the universe not the largest of them all, be the same? Maybe at some point in our growth, long after the end of this existence and we've finished, we'll not just be the field but will also be the new reality.

One possible idea is that we'll be fully evolved and our role is to assist in the creation of new reality, or new universes. This was the view of the occultist W. E. Butler. He stated "We're going to be the universe's builders with God. We'll become instruments and tools with the help of the eternal as His will reigns supreme in the universes He has created and which He dwells and moves his being and that He is working to restore from their shattered condition. You and I also are privileged to be collaborators with Him as well as with the entire of creation, which is an integral an integral part the work of God."

Parting Thoughts

As you continue to follow the steps you've made for yourself and your boat towards the flow, I would suggest that you schedule a whole day each month to plan, afterward, update your plan, update it the plan, and then refine it.

Remember that life is the vision of the Source and you are a part of the dream. There is a role or roles to fulfill. Before you came to earth, you decided to take the roles and took a vow of solemnity to fulfill them. If you take the time I've suggested that you remember that solemn oath , and the vision you had prior to the that self-realized you was born. Now , you're faced with a choice to make. You can either keep your word on your commitment or you could welsh. If you choose to welsh, you'll be able to see the results at the time that your entire life will have to be reconstructed before your judges, you or your guides. Therefore, you should go to an library, or other tranquil location. Spend time thinking, and think about what you want to do.

Making it happen may not be straightforward. It's going to take some the courage. In order to begin it is necessary to clear all the junk out of your mental closet. This is why it is important to take time to work at this without interruptions. It is also important to take time to accept the forgiveness of others and yourself. It is important to overcome your anxieties in order to overcome them and substitute them for positive beliefs. You have to learn to believe in yourself. You have to commit to changes. You must be prepared to go through the hardships. You must let go of your "certainties" of your life at this moment within your own life.

When you've established a strategy that you are following, you must follow the plan. By dedicating half an hour every day, to meditation and every month, a full day to keep an eye on and review your plan. Between, you should be paying attention to your quiet voice. When you look back on your life, you'll be thinking about the decisions that led you to this point. Did they make the right choices? Did you feel satisfied when you made them?

Be sure to find out why your soul was drawn to the circumstances in which you were born. You'll think about your favorite things to do as a kid. You'll ask for help and receive it. You will be in tune with your inner self.

Build trust in your gut instincts by following the advice you get when making small-scale decisions. Then you'll be able to listen to the soft voice you become familiar with when you take the major decisions. It can be scary initially. It can be scary because you don't know whether it's actually the right path to take you to your final destination. However, after a time when you've learned the art of trusting that you are in the right place, uncertainty will become part of the adventure, similar to opening presents at Christmas. You'll be on an adventure that is as thrilling as the ones portrayed in the film Indiana Jones. You will play the role of the director of your own dream.

You could finish the book, put it aside to forget all about it. That's what people do. They've done the things that others have told them to do. They've made a space for themselves. It's not all that thrilling or satisfying however, it is challenging. They've gotten used to the person they are and what they do. What's the reason for this? There's no evidence that they're at risk of becoming further away from their souls and their source that they will never be able to return. It is not possible to establish definitive proof established that a non-physical world exists. A single scientific study has not proven that any aspect that is human is able to survive death. The ones who passed away and were then revived? There are many scientists who believe that it was all within their brains. It's a trick of the brain or a lack of oxygen. In addition, it's a lot of hassle to change. And what do people consider? It's a comfortable life as they are. It's not that bad. What's the point of causing trouble? This Martin person states that once someone starts to make progress, he or will not be able to stop and might end up changing their entire lifestyle like if only they were looking to renovate the kitchen, and ended having to rebuild the entire house.

Sun room done? What is the Den? You'll have to build a wing off towards the rear. Also, a master bedroom on the second floor with skylights and an open fireplace.

While this work is in progress the dust and dirt are piling up and the owner of the home must be amidst the mess.

"Wait for a second I lived in a lovely small bungalow" you might think. "All I needed was a brand new kitchen. It's now turning into an estate. What time will the project be completed?"

The architect, or your higher self will look at you and declare, "Not for a very long period of time I'm scared. You'll need to become familiar with the situation."

Perhaps you're not interested in an estate. If that's how you're feeling, I doubt there's anything I can do right now to make you reconsider your decision. It's probably best to stay to the bungalow.

I'll share a story that I have learned from my own experience. There is nothing more rewarding in this world than doing what you're here for. The process of getting there can be challenging but it's the truth. However, if you are patient and persist and adhere to the steps laid out and receive support, you will be able to get help. In time, you'll begin to see invisible hands that are guiding you, and the path will become easier to locate. The difficulties won't be so difficult to endure. There are blind alleys obviously. There will be failures. There will be difficult lessons to be learned But, over time, you'll arrive at a deep awareness of what your existence as an individual is all about. You will have an intuitive understanding of your place in the overall scheme. You will feel comfortable with the whole thing and still retain your identity. You will begin to understand the things you're doing. You will notice the results manifesting in advance of when they will be. You can decide what you want you want to do and what you should stay clear of.

Once you reach this point, you'll be aware that you've reached a point of power, spiritual power. With this realization, you will feel happiness. Do you think of the joy you'll feel? It could be the mastery of an athletic endeavor like tennis or the mastery in the card bridge game, a musical instrument, or even a foreign language, moment of knowing exactly what you're doing makes you smile.

And also abundance. It's not in the way of money in the traditional sense but the real wealth of the universe that will flow easily to you since you're in the universe, not running upstream.

and health. Your body will react to the new you that you've discovered. There will no longer an excuse for the pain and aches. There is no longer any reason or thought to consider the possibility of dying, because you are on the road to Eternal Life. Your body will become a vibrant living, and thriving cell within the human body and will fulfill your mission and your commitment. Your growth will continue to increase each day and assist others to accomplish the same.

But with this will be a feeling of profound humility as you'll realize that it was not you that led you to this point. It's your subconscious mind, and God's mind. Maybe there is a slight satisfaction realizing that you have learnt to listen. However, you'll need to avoid feeling an overwhelming sense of pride in any way. One of the lessons you'll have learned along the way is that support can be withdrawn from those who believe that they can do great things by themselves. The old saying, "Pride goes before a falls," is certainly true. The people who are proud eventually realize the limits of what they can achieve by themselves.

There will also be an impression of being alone. It's not loneliness because there will be acquaintances, familymembers, and you will have other people with you in the journey. However, very few, if any will have made it to where you are. Very few will be among the people who you can talk to about your experiences and thoughts. They are not all that clear. If you'd like to get a feel of the way this happens go through the Gospels. Each time, you'll see the anger Jesus felt. Sometimes, even his most trusted disciples could not comprehend the meaning behind Jesus' words. Jesus realized that he was one with the Source He named"Father. "Father," and everyone around him believed they were different, distinct and robotic. This is what the majority of people believe they are today.

However you will have an understanding that is new and profound of your worth. It is not possible to imagine yourself as unimportant or meaningless after you have taken into consideration knowing that your self and Source have a common purpose, God is there for you and that guides are always there with you in fact, you're the Force of Life itself seeking self-expression, self-awareness and knowing.

To conclude, let me declare that my desire to you, my prayers, and my desire is that you apply your talents and abilities to benefit the whole creation. I hope that you'll be able to take the opportunity to explore whenever it is offered and join me in the world that Jesus declared to be God's Kingdom. I guarantee you that you won't regret when you realize what you chose to do so.

I feel a sense of gratitude for you, and I am happy in the possibility of having been the conduit to bring you an understanding of you. You are aware of who is your identity, that you're here to achieve a particular purpose, and that your possibilities are endless and limitless. My sincere hope for you is that all my effort been a helping hand to you get another mile along the way. Thanks for taking the one mile journey with me.

Keep going forward. Be sure to always be looking for the bright light. Be prepared for it to be there as it is. Do not hesitate to try it.

Remember that if you believe in yourself that you can, it will be for you.

www.ingramcontent.com/pod-product-compliance
Lightning Source LLC
Chambersburg PA
CBHW060331030426
42336CB00011B/1295